Learning Materials in Biosciences

Peter Pietschmann
Editor

Principles of Bone and Joint Research

Editor
Peter Pietschmann
Department of Pathophysiology and Allergy
Research
Medical University of Vienna
Vienna, Austria

ISSN 2509-6125   ISSN 2509-6133   (electronic)
Learning Materials in Biosciences
ISBN 978-3-319-58954-1   ISBN 978-3-319-58955-8   (eBook)
DOI 10.1007/978-3-319-58955-8

Library of Congress Control Number: 2017950072

© Springer International Publishing AG 2017
This work is subject to copyright. All rights are reserved by the Publisher, whether the whole or part of the material is concerned, specifically the rights of translation, reprinting, reuse of illustrations, recitation, broadcasting, reproduction on microfilms or in any other physical way, and transmission or information storage and retrieval, electronic adaptation, computer software, or by similar or dissimilar methodology now known or hereafter developed.
The use of general descriptive names, registered names, trademarks, service marks, etc. in this publication does not imply, even in the absence of a specific statement, that such names are exempt from the relevant protective laws and regulations and therefore free for general use.
The publisher, the authors and the editors are safe to assume that the advice and information in this book are believed to be true and accurate at the date of publication. Neither the publisher nor the authors or the editors give a warranty, express or implied, with respect to the material contained herein or for any errors or omissions that may have been made. The publisher remains neutral with regard to jurisdictional claims in published maps and institutional affiliations.

Printed on acid-free paper

This Springer imprint is published by Springer Nature
The registered company is Springer International Publishing AG
The registered company address is: Gewerbestrasse 11, 6330 Cham, Switzerland

Preface

Bone and joint diseases (such as osteoporosis or osteoarthritis) are among the leading causes of morbidity and confer a significant burden on individuals and societies. Inspired by the «Bone and Joint Decade 2000–2010,» the Medical University of Vienna established the doctoral program «Regeneration of Bones and Joints.» The objective of this program is to train doctoral students to become the next generation of advanced researchers in musculoskeletal and oral sciences. In a translational and multidisciplinary approach, pathophysiologic mechanisms and novel therapeutic strategies for bone, joint, and dental diseases are investigated. The curriculum also includes lectures, seminars, and practical trainings.

This book is intended as an accompanying book for the program «Regeneration of Bones and Joints» of the Medical University of Vienna and similar doctoral or master's programs. It describes basic principles and methods of bone and joint research. In addition, the pathophysiology of selected musculoskeletal and oral diseases is presented.

I am very thankful to the authors of this book for their effort, time, and expertise. I would also like to acknowledge the important help and dedication of the Springer Nature staff, in particular Dr. Silvia Herold. Finally I, wish to thank Birgit Schwarz for her continuous support of this book project.

I am convinced that this book will contribute to advance our knowledge in musculoskeletal and oral sciences.

Peter Pietschmann
Vienna, Austria

Contents

1	**Bone Material Quality**	1
1.1	Structure and Composition of Bone at Material Level	2
1.1.1	Bone Structural Units (BSUs)	2
1.1.2	Lamellar Arrangement of Collagen Fibrils	4
1.1.3	Mineralized Collagen Fibrils	5
1.1.4	Collagen and Mineral	5
1.2	Quality and Mechanical Properties of Bone Material	6
1.2.1	From Mechanical Properties of Whole Bone to Properties of Bone Material	6
1.2.2	Bone Material Quality	7
1.3	Overview of Methods for Characterization of Bone Material	8
1.4	Bone Fragility Associated with Altered Bone Material	12
1.4.1	Bone Material in Postmenopausal Osteoporosis and After Treatment	12
1.4.2	Bone Material in Osteogenesis Imperfecta	12
	References	13
2	**Osteoblasts and Osteocytes: Essentials and Methods**	17
2.1	From Mesenchymal Stem Cells to Osteocytes	18
2.2	Inverse Relationship Between Osteogenic and Adipogenic Programming	19
2.3	From Osteoblasts to Osteocytes	22
2.4	The Osteocyte	22
2.5	Some Essential Facts About Osteocytes	23
2.6	Conclusion	25
2.7	Experimental Systems to Study Osteoblastic Differentiation and Behavior	25
	References	30
3	**Osteoclasts: Essentials and Methods**	33
3.1	Osteoclast Biology	34
3.1.1	Osteoclastogenesis	34
3.1.2	Osteoclast Functions	36
3.2	Osteoclast Deregulation and Related Pathologies	39
3.2.1	Osteoporosis	39
3.2.2	Bone Metastases	40
3.2.3	Osteopetrosis	41
3.3	Osteoclast Pharmacology	42
3.3.1	Bisphosphonates	43
3.3.2	Denosumab	43
3.3.3	New Antiresorptive Agents	44
3.4	Methods for Osteoclast Primary Cultures	44
3.4.1	Osteoclast Isolation from Mouse Bone Marrow Cells	45
3.4.2	Osteoclast Isolation from Human Peripheral Blood	45
3.4.3	Evaluation of Osteoclast Differentiation	46
3.4.4	Evaluation of Osteoclast Function	47
3.5	Conclusions	47
	References	48

4	**Bone Turnover Markers**	55
4.1	Introduction	57
4.2	**Bone Resorption Markers**	58
4.2.1	Hydroxyproline	58
4.2.2	Pyridinium Cross-Links	58
4.2.3	Telopeptides of Collagen Type 1	58
4.2.4	Tartrate-Resistant Acid Phosphatase Type 5	59
4.3	**Bone Formation Markers**	59
4.3.1	Propeptides of Type 1 Collagen	59
4.3.2	Bone-Specific Alkaline Phosphatase	59
4.3.3	Osteocalcin	60
4.4	**Animal Studies**	60
4.5	**Further Bone Turnover Markers**	60
4.5.1	RANK/RANKL/OPG System	61
4.5.2	Cathepsin K	61
4.5.3	Periostin	61
4.5.4	Dickkopf 1	61
4.5.5	Sclerostin	61
4.5.6	Fibroblast Growth Factor 23	62
4.5.7	MicroRNAs	62
4.6	**Non-pathological and Pathological Conditions Leading to Changes in BTMs**	62
4.6.1	Bone Turnover Markers in Patients with Chronic Kidney Disease	62
4.7	**Changes in Established Bone Turnover Markers with Osteoporosis Treatment**	63
4.8	**Clinical Relevance of Bone Turnover Markers**	63
	References	64
5	**Bone Imaging by Nuclear Medicine**	67
5.1	Introduction	68
5.2	**Bone Scintigraphy: General Information**	68
5.3	**Bone Disorders with Pathological Bone Scintigraphies**	70
5.3.1	Primary Bone Tumors	70
5.3.2	Bone Metastases	72
5.4	**Metabolic Bone Disorders**	74
5.4.1	Osteoporosis	74
5.4.2	Osteomalacia	75
5.4.3	Primary Hyperparathyroidism	75
5.4.4	Renal Osteodystrophy	77
5.4.5	Hyperthyroidism	77
5.4.6	Paget's Disease of Bone	77
5.4.7	Fibrous Dysplasia	77
5.5	**Inflammation**	78
5.5.1	Joint Inflammation	78
5.5.2	Chronic Osteomyelitis	78
5.5.3	Degenerative Bone and Joint Changes	79
5.5.4	Algodystrophy	80
5.6	**Conclusion**	82
	References	82

6	**Pathophysiology of Bone Fragility**	83
6.1	**Determinants of Bone Fragility**	84
6.1.1	Bone Mass	84
6.1.2	Macroarchitecture	85
6.1.3	Microarchitecture	85
6.1.4	Material Properties	86
6.1.5	Bone Turnover	86
6.1.6	Bone Fragility: Concluding Remarks	87
6.2	**Pathophysiology of Osteoporosis**	87
6.2.1	Definition and Risk Factors	87
6.2.2	Types of Osteoporosis	88
6.2.3	Pathophysiology	88
6.2.4	Postmenopausal Bone Loss	90
6.2.5	«Senile» Bone Loss	90
6.3	**Pathophysiology of Non-osteoporotic Bone Diseases**	91
6.3.1	Paget's Disease of Bone	91
6.3.2	Primary Hyperparathyroidism	92
6.3.3	Rickets and Osteomalacia: Disorders of Bone Mineralization	92
6.3.4	Chronic Kidney Disease-Mineral Bone Disorder	94
	References	95
7	**Fracture Healing**	99
7.1	**Introduction**	100
7.2	**Etiology**	101
7.3	**Epidemiology**	102
7.4	**Principles of Fracture Treatment/Osteosynthesis**	102
7.5	**Experimental Models in Fracture Repair**	104
7.6	**Physiology of Fracture Healing**	108
7.6.1	Primary/Direct (Intramembranous) Healing	108
7.6.2	Secondary/Indirect (Endosteal/Intramembranous) Healing	109
7.7	**Pathophysiology of Fracture Healing**	116
7.7.1	Classifications of Nonunions (NU)	117
7.7.2	Molecular Alterations	118
7.7.3	Treatment Modalities	120
	References	123
8	**Oral and Maxillofacial Aspects of Bone Research**	125
8.1	**Development of the Facial Skeleton**	126
8.2	**Periodontitis: Defense Mechanisms Versus Bacterial Invaders**	127
8.3	**Medication-Related Osteonecrosis of the Jaws: The Phossy Jaw of the Twenty-First Century**	129
8.4	**Bone Morphogenetic Proteins: The Bone Scientist's Alchemic Compound**	131
	References	134
9	**Cartilage Biology: Essentials and Methods**	139
9.1	**Introduction**	140
9.2	**Articular Cartilage Function**	141
9.3	**Articular Cartilage Morphology**	141

9.4	Subdivision of Articular Cartilage in Four Layers	143
9.5	Compartmentalisation of the ECM in Relation to the Chondrocytes	147
9.6	The Chondrocyte and its Pericellular Microenvironment	148
9.7	Articular Cartilage Structure Under Load	149
9.8	Morphological Methods for Cartilage Research	150
	Suggested Reading	150
10	**Growth Plate Research**	**153**
10.1	Introduction	154
10.2	GP Structure and Function	154
10.2.1	Endochondral Ossification	154
10.2.2	GP Structure and Function Is Regulated by Distinct Steps of Differentiation and Proliferation	155
10.2.3	Complex Local Regulation Loops Regulate Proliferation, Differentiation, and Height of the GP	156
10.2.4	Endocrine Regulation of Growth Is Dominated by the GH-IGF-I Pathway	157
10.2.5	The Mechanisms by Which Estrogen Regulates GP Maturation and Fusion Is Still Not Fully Understood	158
10.3	Techniques in GP Research	158
10.3.1	Histology	158
10.3.2	Immunohistochemistry	160
10.3.3	Electron Microscopy	161
10.3.4	Laser Capture Microdissection (LCM)	163
10.3.5	Transgenic Animal Models in GP Research	164
10.3.6	In Vitro Model Systems in GP Research	165
	References	168
11	**Pathogenetic Concepts of Joint Diseases**	**173**
11.1	Joint Affection in Systemic Autoimmune Disorders	174
11.2	Inflammatory Joint Diseases	174
11.2.1	Synovial Inflammation	175
11.2.2	Key Players in Synovial Inflammation	175
11.3	Biologic Treatment Options in Rheumatoid Arthritis	181
11.3.1	TNF Blockade	182
11.3.2	IL-1 Neutralization	182
11.3.3	IL-6 Blockade	182
11.3.4	Co-stimulation Modulation	183
11.3.5	B Cell Depletion	183
	References	184
12	**Osteoarthritis Biology**	**189**
12.1	Osteoarthritis	190
12.1.1	Risk Factors	190
12.1.2	Socioeconomic Impact	191
12.1.3	Pathophysiology of OA	191

12.2	**Triggers of OA Biology**	194
12.2.1	Ageing	194
12.2.2	Obesity	195
12.2.3	Genetics	195
12.2.4	Inflammation	196
12.2.5	Subchondral Bone	198
12.2.6	Signalling	199
12.3	**Concluding Remarks**	201
	References	202

Supplementary Information

Index 207

Contributors

Stéphane Blouin, PhD
Ludwig Boltzmann Institute of Osteology at
the Hanusch Hospital of WGKK and
AUVA Trauma Centre Meidling
1st Medical Department Hanusch Hospital
Vienna, Austria
stephane.blouin@ osteologie.at

Monika Egerbacher, DVM
Institute of Anatomy,
Histology and Embryology
Veterinary University Vienna
Vienna, Austria
monika.egerbacher@vetmeduni.ac.at

Ursula Foeger-Samwald, PhD
Department of Pathophysiology
and Allergy Research,
Center for Pathophysiology,
Infectiology and Immunology
Medical University of Vienna
Vienna, Austria
ursula.foeger-samwald@meduniwien.ac.at

Nadja Fratzl-Zelman, PhD
Ludwig Boltzmann Institute of Osteology
at the Hanusch Hospital of WGKK and
AUVA Trauma Centre Meidling,
1st Medical Department Hanusch Hospital
Vienna, Austria
nadja.fratzl-zelman@osteologie.at

Sonja Gamsjaeger, PhD
Ludwig Boltzmann Institute of Osteology
at the Hanusch Hospital of WGKK
and AUVA Trauma Centre Meidling,
1st Medical Department Hanusch Hospital,
Vienna, Austria
sonja.gamsjaeger@gmx.at

Nicolas Haffner, MD
Orthopedic Hospital Gersthof
Vienna, Austria
n.haffner@osteodoc.at

Gabriele Haeusler, MD
University Clinic of Paediatrics
and Adolescent Medicine,
Medical University of Vienna,
Vienna, Austria
gabriele.haeusler@meduniwien.ac.at

Katharina Kerschan-Schindl, MD
Department of Physical Medicine
Rehabilitation and Occupational Medicine
Medical University of Vienna
Vienna, Austria
katharina.kerschan-schindl@meduniwien.ac.at

Klaus Klaushofer, MD
Ludwig Boltzmann Institute of Osteology at the
Hanusch Hospital of WGKK and AUVA Trauma
Centre Meidling, 1st Medical Department
Hanusch Hospital
Vienna, Austria
klaus.klaushofer@osteologie.at

Jan Leipe, MD
Division of Rheumatology and Clinical
Immunology
Ludwig – Maximilians – University of Munich,
Munich, Germany
jan.leipe@med.uni-muenchen.de

Stefan Marlovits, MD
Department of Trauma Surgery
Medical University of Vienna
Vienna, Austria
stefan.marlovits@meduniwien.ac.at

Peter Mikosch, MD
Department of Internal Medicine 2,
Landesklinikum Mistelbach-Gänserndorf,
Mistelbach, Austria
peter.mikosch@mistelbach.lknoe.at

Contributors

Barbara Misof, PhD
Ludwig Boltzmann Institute of Osteology
UKH Meidling
Kundratstr. 37, 1120 Vienna, Austria
barbara.misof@osteologie.at

Rainer Mittermayr, MD
Ludwig Boltzmann Institute for
Experimental and Clinical Traumatology,
AUVA Trauma Centre Meidling
Vienna, Austria
Rainer.Mittermayr@TRAUMA.LBG.AC.AT

Sylvia Nürnberger, PhD
Department of Trauma Surgery
Medical University of Vienna
Vienna, Austria
sylvia.nuernberger@meduniwien.ac.at

Eleftherios Paschalis, PhD
Ludwig Boltzmann Institute of
Osteology at the Hanusch Hospital
of WGKK and AUVA Trauma Centre Meidling,
1st Medical Department Hanusch Hospital
Vienna, Austria
eleftherios.paschalis@osteologie.at

Peter Pietschmann, MD
Department of Pathophysiology
and Allergy Research,
Center for Pathophysiology,
Infectiology and Immunology
Medical University of Vienna
Vienna, Austria
peter.pietschmann@meduniwien.ac.at

Adalbert Raimann, MD
University Clinic of Paediatrics
and Adolescent Medicine
Medical University of Vienna
Vienna, Austria
adalbert.raimann@meduniwien.ac.at

Paul Roschger, PhD
Ludwig Boltzmann Institute of
Osteology at the Hanusch Hospital
of WGKK and AUVA Trauma Centre Meidling,
1st Medical Department Hanusch Hospital
Vienna, Austria
paul.roschger@osteologie.at

Nadia Rucci, PhD
Department of Biotechnological
and Applied Clinical Sciences
University of L'Aquila
L'Aquila, Italy
nadia.rucci@univaq.it

Hendrik Schulze-Koops, MD
Department of Internal Medicine IV –
Endocrinology, Nephrology, Rheumatology
LMU – University Hospital Munich
Munich, Germany
hendrik.schulze-koops@med.uni-muenchen.de

Daniel Smolen, MD
Ludwig Boltzmann Institute for
Experimental and Clinical Traumatology,
AUVA Trauma Center Meidling
Vienna, Austria
daniel.smolen@icloud.com

Anna Teti, PhD
Department of Biotechnological
and Applied Clinical Sciences
University of L'Aquila
L'Aquila, Italy
annamaria.teti@univaq.it

Stefan Toegel, MD
Karl Chiari Lab for Orthopaedic Biology,
Department of Orthopaedics
Medical University of Vienna
Vienna, Austria
stefan.toegel@meduniwien.ac.at

Franz Varga, PhD
Ludwig Boltzmann Institute of
Osteology at the Hanusch Hospital of
WGKK and AUVA Trauma Centre Meidling,
1st Medical Department Hanusch Hospital
Vienna, Austria
franz.varga@osteologie.at

Arno Wutzl, MD
Zahnarzt, Facharzt für Mund-,
Kiefer- und Gesichtschirurgie, Tulln, Austria
and Trauma Centre Meidling
Vienna, Austria
arno@zahnarztwutzl.at

Bone Material Quality

Paul Roschger, Stéphane Blouin, Eleftherios Paschalis, Sonja Gamsjaeger, Klaus Klaushofer, and Barbara Misof

1.1	Structure and Composition of Bone at Material Level – 2
1.1.1	Bone Structural Units (BSUs) – 2
1.1.2	Lamellar Arrangement of Collagen Fibrils – 4
1.1.3	Mineralized Collagen Fibrils – 5
1.1.4	Collagen and Mineral – 5
1.2	Quality and Mechanical Properties of Bone Material – 6
1.2.1	From Mechanical Properties of Whole Bone to Properties of Bone Material – 6
1.2.2	Bone Material Quality – 7
1.3	Overview of Methods for Characterization of Bone Material – 8
1.4	Bone Fragility Associated with Altered Bone Material – 12
1.4.1	Bone Material in Postmenopausal Osteoporosis and After Treatment – 12
1.4.2	Bone Material in Osteogenesis Imperfecta – 12
	References – 13

© Springer International Publishing AG 2017
P. Pietschmann (ed.), *Principles of Bone and Joint Research*, Learning Materials in Biosciences,
DOI 10.1007/978-3-319-58955-8_1

What You Will Learn in This Chapter

One important role of our skeleton is to fulfil mechanical functions, in particular the mechanical support of our body enabling complex movement and locomotion as well as safety housing for sensitive organs. In this book chapter, you will learn about the characteristics of the bone material and their role for the mechanical performance of the bone at organ level. First, we give an overview of the hierarchical organization of the bone, the heterogeneity of bone material structure and the composition from microscale to nanoscale. Second, we describe important technical terms related to bone deformation; we present bone material quality parameters and discuss their role in bone deformation. Third, we briefly explain the most important current techniques for the characterization of the bone material quality such as light and electron microscopy methods, X-ray scattering and fluorescence, vibrational spectroscopic techniques, etc. Forth, we present material characteristics in clinical examples of bone fragility including postmenopausal osteoporosis and treatment and osteogenesis imperfecta.

1.1 Structure and Composition of Bone at Material Level

Bone is a *lightweight structured* organ, i.e. it is constructed following the principle to have maximum resistance to fracture based on as little material/mass as possible. For this purpose, bone is hierarchically organized [1]. One may distinguish between organ, tissue and material levels (◘ Fig. 1.1). The *organ level* (outer geometry, shape and dimensions of the whole bone) and the *tissue level* (inner architecture of the bone) and their role in bone fragility are described in the book ▶ Chap. 6 of Pietschmann et al. In general, all hierarchical levels are optimized for the excellent mechanical competence of the bone [2]. As discussed later, bone diseases associated with fragility can be caused by either affecting single or multiple hierarchical levels of bone organization.

1.1.1 Bone Structural Units (BSUs)

The ongoing process of bone resorption and formation (bone remodelling) is responsible for a remarkable heterogeneity of the bone material. During bone remodelling, bone formation is occurring on surfaces which were generated by previous resorption due to osteoclastic activity (see book ▶ Chap. 3 of Teti and Rucci) and covered by a thin non-collagenous protein cement layer. The bone material which is deposited on a pre-existing bone surface by the coordinated action of numerous osteoblasts (see book ▶ Chap. 2 of Fratzl-Zelman and Varga) during formation time is called a bone structural unit (BSU) (in trabecular bone also termed «bone packet» or «hemiosteon», in cortical bone «osteon»). Depending on the time passed from formation, each BSU has its specific tissue age. The bone turnover rate (frequency of remodelling cycles) determines the distribution and the average of these BSUs' tissue ages.

An important feature of a BSU is its mineral content, the bone matrix mineralization density. When bone matrix is deposited by the osteoblasts, its mineral accumulation follows a specific time course. First, during the osteoid phase, the collagenous matrix is maturing without being mineralized. Second, after onset of mineralization, mineral accumulates rapidly during the primary mineralization. This is followed by a third period with slowly increasing matrix mineralization (secondary mineralization) which takes

Bone Material Quality

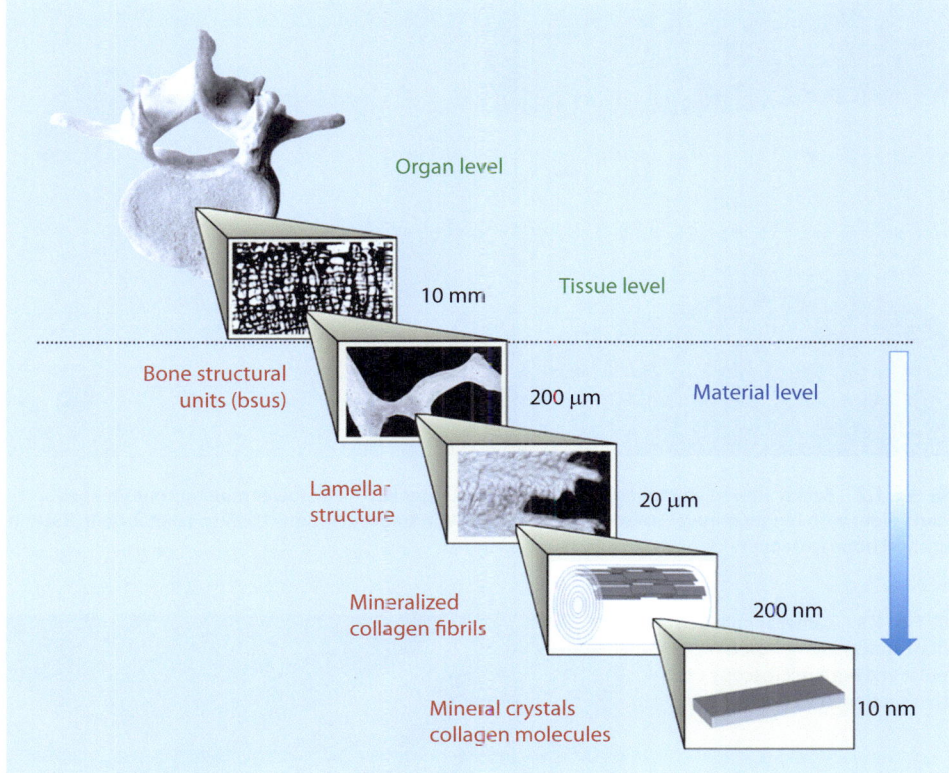

Fig. 1.1 Hierarchical structure of the bone

months to years until the plateau of full mineralization is achieved [3, 4]. Due to the time course of these processes, the bone is composed predominantly of BSUs in the secondary mineralization phase. In healthy individuals, the global mineralization pattern reflects the tissue ages of the BSUs (younger bone tissue is less, older higher mineralized) (◘ Fig. 1.2).

Another characteristic feature of a BSU is its lamellar collagen fibril arrangement (the next lower level of bone material organization; see ▶ Sect. 1.2). In trabecular bone a uniform lamellar structure parallel to the trabecular surface is formed, while in cortical bone, the lamellar bone is arranged cylindrically around a Haversian canal (a canal for blood vessels, nerves and bone marrow). In general, this orientation is abruptly changing at the cement line, which represents the demarcation to the adjacent BSU (◘ Fig. 1.3).

When the osteoblasts deposit the matrix, part of these cells gets entrapped in the mineralized bone matrix and differentiates into osteocytes which form a dense osteocyte lacunar-canalicular network [5] with a density of about 20,000 osteocyte lacunae per mm^3 [6] and an average osteonal canalicular density of $1.6 \pm 0.8 \times 10^6/mm^3$ [7]. There is evidence that this network is involved in mechanosensing of load by local bone deformation and in cell communication. Moreover, this lacuno-canalicular system provides an enormous inner bone surface of about 215 m^2 [8] (see book ▶ Chap. 2 of Fratzl-Zelman and Varga), which might be used to regulate rapidly the homeostasis of elements like Ca, P and Mg.

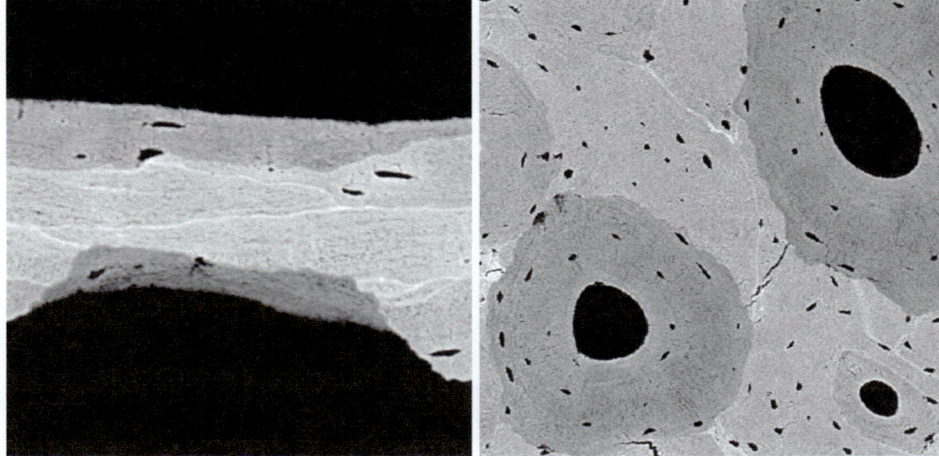

Fig. 1.2 Backscattered electron images of BSUs. Younger BSUs with lower mineral content appear dark, older with higher mineral content bright; *left*, BSUs in trabecular bone (bone packets); *right*, BSUs in cortical bone (osteons)

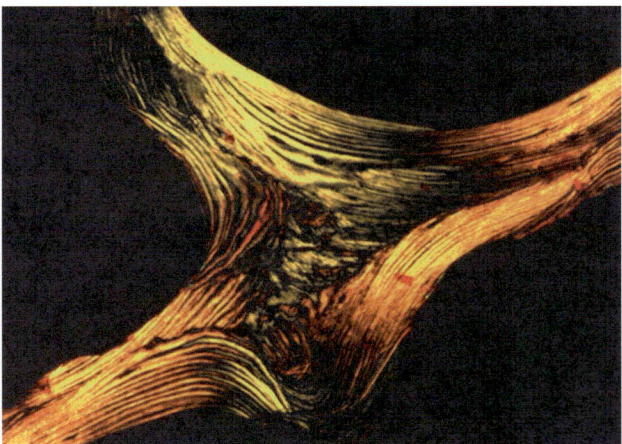

Fig. 1.3 Lamellar collagen fibril arrangement based on polarized light microscopy of a thin section of trabecular bone

1.1.2 Lamellar Arrangement of Collagen Fibrils

In mature bone, the collagen fibrils are arranged in parallel bundles, which are embedded in an extra-fibrillar matrix (organic and mineral) [9]. They are further organized to form a lamellar structure with a period of about 7 μm produced by a periodic change of the dominant fibril orientation (this is in contrast to primary/callus bone, where the fibrils are oriented randomly) [10]. A narrow and a wide lamella can be distinguished. In osteonal bone, a helicoidal structure of the collagen fibrils arranged in a spiral around the blood vessel can be observed [10].

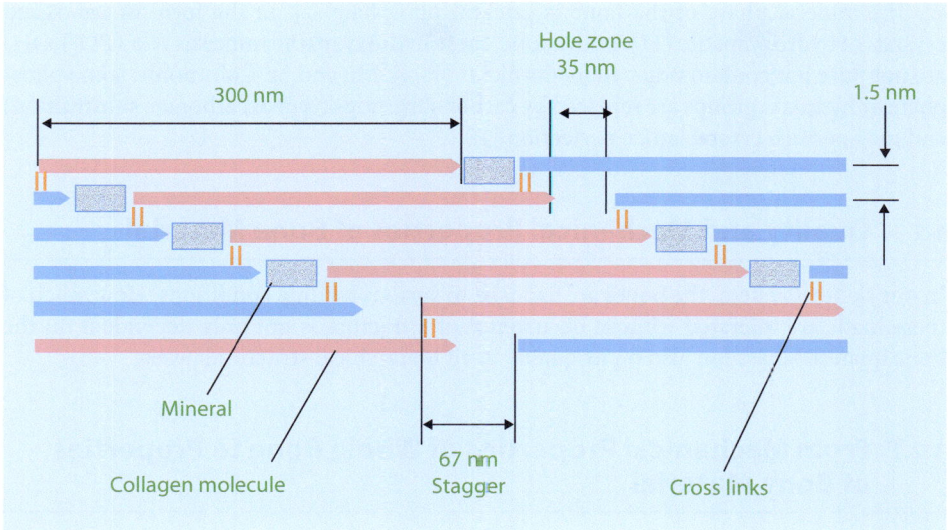

Fig. 1.4 Schematic drawing of collagen fibril with starting mineralization. The colours of the collagen fibrils indicate the periodicity in their arrangement

1.1.3 Mineralized Collagen Fibrils

The mineralized collagen fibril is the basic building block of the bone material [11]. Within the fibril, about 300 nm long molecules line up parallel staggered and form a characteristic banded structure with a 67 nm periodicity of hole and overlap zones of collagen molecules [12] (◘ Fig. 1.4). It is assumed that during the mineralization process (see ▸ Sect. 1.1), the mineral crystals start growing predominantly in the hole zones of the fibril to form plate-like mineral particles of 1.5 to 4.5 nm thickness, about 50 nm length and 25 nm width [13, 14]. They are also arranged in a parallel, staggered manner with orientation to the longitudinal axis of the fibrils [2].

1.1.4 Collagen and Mineral

The components of the nanocomposite bone material are *organic molecules* and *mineral crystals*. The main organic part (90%) is collagen type I, and the rest (10%) are non-collagenous proteins which are both important for initiation of mineralization [15, 16]. After the secretion of procollagen molecules by the osteoblasts, the collagen fibrils form by self-assembly and get mineralized later (see ▸ Sect. 1.1). Enzymatic and non-enzymatic cross-links are formed between the collagen helices to stabilize the collagen fibrils. Divalent cross-links are first formed by enzymatic involvement (lysyl oxidase) partly followed by transformation to trivalent cross-links spontaneously [2].

The mineral phase of the bone is calcium phosphate salt in the form of nanosized crystals of hydroxyapatite [17]. In the bone, these hydroxyapatite minerals ($Ca_5(PO_4)_3OH$) are not pure but contain other elements like F, Na, K, Mg and Sr. Commonly, some phosphate or hydroxyl groups are replaced by carbonate groups (type B carbonate substitution) and may perturb crystal lattice perfection [18].

1.2 Quality and Mechanical Properties of Bone Material

In our skeletal system, the bones are exposed to various loading conditions. How much of these loads are tolerated without occurrence of a fracture is not only dependent on the bone quantity/mass but also on its quality at all hierarchical structural levels.

1.2.1 From Mechanical Properties of Whole Bone to Properties of Bone Material

The bones of our skeletal system can experience three principle loading modes: tension, compression and shear and the combination thereof. Estimation of whole-bone mechanical properties is typically based on compression tests (e.g. for vertebrae) or three- and four-point bending tests (for long bones); for details see reference [19]. During a typical compression test, e.g. both the progress of loading (increase of force) with time and the corresponding deformation (displacement of the vertebral endplate) are recorded in the load-displacement curve. Characteristic mechanical indices derived from this curve (◘ Fig. 1.5) are the applied load (y-axis), the resulting displacement (x-axis), the linear part of the curve (the elastic region), the slope of the elastic region (stiffness), the non-linear region (indicating the plastic region, where deformation is irreversible), the displacement and load (failure load) at which bone fractures and the entire area under the curve which represents the work to failure. Furthermore, single loading (increasing load), cyclic loading (repetitive lower magnitude loading) or creep experiments (constant load) can be distinguished but will not be further described here.

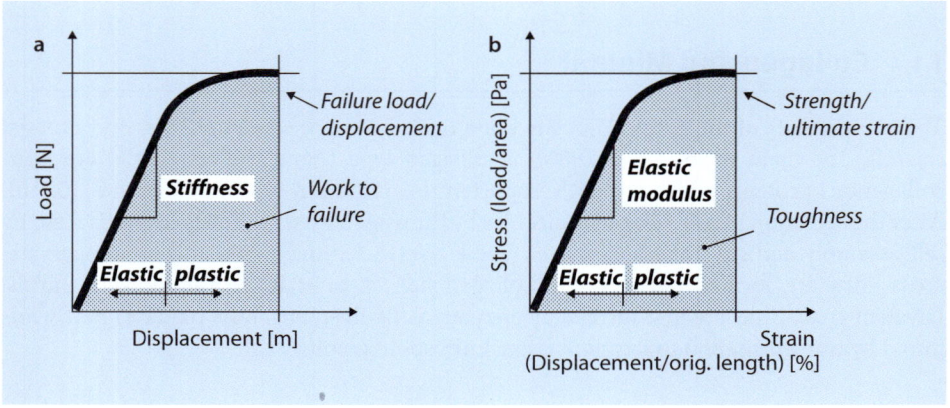

◘ Fig. 1.5 Characteristic mechanical indices derived from **a** load-displacement or **b** stress-strain curve

Such mechanical tests on whole bone or even on small tissue samples are dependent on both geometrical/structural and material properties. For instance, an important geometrical parameter of the long bone is its cross-sectional outer radius, as the bending stiffness of the bone increases with a power 4 law of its outer radius. Moreover, cortical thickness, porosity, density and size of Haversian canals as well as trabecular architecture (see also book ▶ Chap. 6 of Pietschmann et al.) play an important role for whole-bone mechanical properties. In general, if sufficient geometrical information is available, *indirect estimates* of mechanical properties of material (in contrast to the *direct measures* in ▶ Sect. 1.3, where an overview of methods for bone material characterization is given) can be deduced from such experiments. Subsequently, the load-displacement curves normalized for bone/sample geometry reveal indirect estimates of bone material such as stress, strain, E-modulus, strength and toughness (◘ Fig. 1.5).

1.2.2 Bone Material Quality

Apart from macroscopic changes, deformation and failure at multiple lower size scales (at material level) play a role in the deformation of the bone as a whole.

Role of Mineralization Pattern for Deformation

The slope of the pre-yield (elastic) region in the stress-strain curve was shown to be related to the degree of matrix mineralization [20]. The higher the degree of mineralization or density of the mineral particles, the larger is the slope (the stiffer is the material) [21]. Both hypo- and hypermineralization may compromise the mechanical performance of the bone [22]. Our previous findings suggest an optimal bone mineralization density distribution throughout the bone matrix [23]. Another important parameter of mineralization is its heterogeneity (introduced by the BSUs) that may play a role for crack propagation and deflection as discussed below.

Role of the Microcomposite for Deformation

The combination of soft and tough collagen with the hard and brittle mineral leads to an extraordinary composite which is stiff but not brittle. The special arrangement and the quality of the two components are essential as tensile stress on the bone material is transmitted through the collagenous matrix by shear to the crystals protecting them from fracture [9, 24]. Abnormal tendon collagen in a mouse model of osteogenesis imperfecta was found to have decreased yield and ultimate stress and strain as well as decreased toughness [25]. Abnormal enzymatic and non-enzymatic cross-linking was related to increased bone fragility [26–30]. Additionally, deviations from normal mineral crystal shape and size as observed after sodium fluoride treatment were linked to decreased bone material strength [31]; see ▶ Sect. 4.1.

For the direct observation of *deformation at nanoscale*, sophisticated techniques including in situ experiments at the synchrotron are required which enable simultaneous mechanical testing and high-resolution imaging. Approximately 58% of the tissue strain in the elastic region (overall deformation) during tensile testing occurs via shearing within the extra-fibrillar matrix (between the fibrils) and 42% within the mineralized fibrils. The mineral platelets receive only 16% of the overall deformation. This indicates a natural stress shielding design where the brittle mineral particles see less strain to avoid catastrophic failure. When deformation in these tensile tests exceeded the elastic region, fibrils showed no further elongation suggesting the slippage of the fibrils [9].

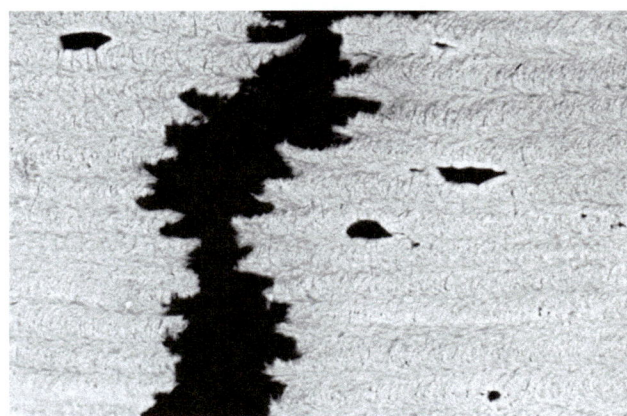

Fig. 1.6 Backscattered electron image of crack propagation perpendicular to the lamellar orientation of the bone material. «Zig-zag» line indicates the requirement of much energy for propagation

Toughening Mechanisms of Bone Material

When the bone is loaded beyond the elastic region, damage and cracks (deformation) become apparent at *microscopic scale*. It was observed that the formation of such microcracks and/or diffuse microdamage increased bone toughness, thus being a mechanism of energy absorption. The underlying mechanisms were proposed to be (i) crack (or «ligament») bridging and (ii) crack deflection [32]. Indeed, lamellar orientation plays an important role, and it was reported that it requires much higher energies to propagate a crack perpendicular than parallel to the lamellar orientation [33] (◘ Fig. 1.6). Disturbed lamellar orientation was linked to bone fragility in pycnodysostosis [34]. Heterogeneity of bone material in general (interfaces with changing mineral content, lamellar orientation, organic matrix) might be important for crack deviation and crack stopping [35–38]. There is also evidence that the osteocyte lacunar-canalicular network plays a role in crack initiation and propagation [7].

Although the formation of microcracks might enhance toughness, once they have been formed, their presence might weaken the bone material in response to future external loading [39]. Removal of microdamage has been proposed to be essential and to be conducted by targeted resorption. Microdamage involves the damage of the osteocytic network, and signals from the osteocytes have been proposed to attract the osteoclasts to these damaged sites for resorption and removal of damaged bone material. This targeted bone resorption is generally assumed to be an important bone remodelling mechanism.

1.3 Overview of Methods for Characterization of Bone Material

During recent years, several methods and the combination thereof have been developed which allow comprehensive characterization of structural and mechanical properties at high spatial resolution. We give here an overview of selected methods and their application on bone material; for more information, see reference [40].

- **Arrangement and Composition of Organic Matrix**

Polarized light microscopy: The lamellar arrangement of the collagen fibrils can be visualized due to the birefringence of collagen.

Fourier transform infrared (FTIR) and Raman microspectroscopy (RM): Vibrational spectroscopic methods are sensitive to bond vibrations thus providing information on the functional groups present in the mineral and organic matrix components, as well as the short- and medium-range interactions between them («molecular neighbourhood»). In FTIR thin (~4 μm) sections are required, whereas in RM, sections and/or bone blocks are analysed. Both techniques allow the characterization of the relative amounts of organic matrix and mineral, information on carbonate substitution in the mineral, mineral crystallinity/maturation and collagen cross-linking. RM has been recently established also for the measurement of lipids, proteoglycans and advanced glycation end products (AGEs) [41, 42]. Recently, a new parameter has been introduced which provides information on the nanoporosity (a surrogate for tissue water) of bone sample. During sample preparation, the embedding substrate penetrates the porosities of the bone material which had been previously (in-vivo) filled with water. In this way, the relative amount of the embedding material can be used for an estimation of the nanoporosity/water content within the bone material [43].

- **Quantification of Mineral Content Distribution**

Quantitative Backscattered Electron Imaging (qBEI): qBEI in the scanning electron microscope is a sensitive and frequently applied technique for the analysis of the mineralization pattern [23]. It requires block samples of undecalcified polymethylmethacrylate embedded tissue with a flat surface. qBEI is based on the dependency of the backscattered electron intensity on the local atomic number in the sample (mainly dependent on the local calcium concentration). The higher the calcium concentration, the higher the backscattered electron intensities and the brighter are the pixel grey levels in the image. The images can be used for the analysis of histograms revealing the frequency of the different calcium concentrations (given in weight% Ca) in the studies bone sample (designated bone mineralization density distribution, BMDD), ◘ Fig. 1.7.

Quantitative Microradiography (qMR): This was the first method used to quantify the mineralization pattern. qMR is based on the unidirectional X-ray projection of 100 μm thick polymethylmethacrylate embedded bone sections and the measurement of the absorption of the former. The resulting microradiograph is analysed for its grey levels and grey level histograms having the meaning of mineral content in units of g/cm^3 [44].

Synchrotron Radiation Microtomography (SRμCT): SRμCT is the most modern method and provides 3-dimensional information on the mineral content in the bone sample. It is based on multidirectional projection and absorption of a focused, monochromatic X-ray beam by small bone samples (2 mm × 2 mm). The resulting data set allows 3-D reconstruction and mineral distribution measurement based on voxel grey levels with the meaning of mineral content in units of g/cm^3 [40].

- **Characteristics of Mineral Particles/Crystals**

Scanning Small-Angle X-Ray Scattering (sSAXS): This method is based on the measurement of the scattered intensities very close to the incident beam direction (at small angles). In bone material analysis, this method is mainly used for information on the size distribution, the shape and the arrangement of the mineral particles of the nanocomposite material with size dimensions smaller than 50 nm. sSAXS is available at the laboratory or at the synchrotron (the latter allows to scan the sample with high spatial resolution of 3–10 μm; however, it requires very thin 3–10 μm bone sections) [14, 31]. Recently, sSAXS tomography was introduced which provides the 3D view of mineral particle properties of the bone [45].

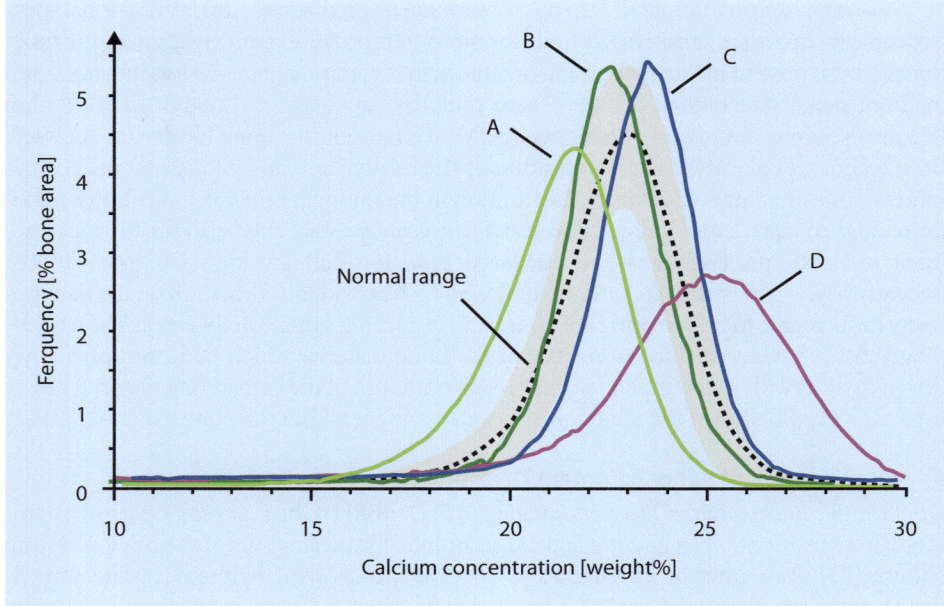

Fig. 1.7 Examples of bone mineralization density distribution in health and disease. *A* postmenopausal osteoporosis, *B* bisphosphonate treated, *C* osteogenesis imperfecta, *D* sodium fluoride treated bone

Scanning Wide-Angle X-Ray Scattering (sWAXS): The scattered X-ray intensities under larger (wide) angles can originate from the scattering/diffraction of the incident X-ray beam by the crystal lattice of the bone mineral particles and provide information on crystal lattice characteristics (lattice parameters of the crystal unit cell, crystallinity and crystal dimensions) [10].

Transmission Electron Microscopy (TEM): Another method to study the small bone mineral particles is TEM, which provides direct high-resolution images or might be used for electron microscopic tomography [40]. For TEM analysis, plastic embedded, ultra-thin bone sections (70–100 nm) are required. The energy of the electrons, which have to transmit the bone section, is in the range of 100–1000 keV. The image contrast is generated mainly by the difference in electron density of the structures in the sample. Single/isolated crystals in the viewing field can be analysed for their shape and dimensions.

Elemental Composition

X-Ray Fluorescence (ERF) Induced by Electrons or by X-Ray at the Synchrotron (SR-XRF): These techniques allow non-destructively, spatially resolved analysis of the elemental composition of the bone (Ca, P, Mg, Na, K). SR-XRF is applied when sufficient sensitivity (ppm concentration range) is needed, e.g. in the detection of trace elements (Pb, Sr, Mn, Zn, F) [46].

Visualization and Quantification of Osteocyte Lacunar-Canalicular Network (OLCN): The OLCN in the bone material can be studied in three dimensions by confocal laser scanning microscopy (CLSM) after fluorescent staining of bulk bone samples with rhodamine G, e.g. which stains the organic phase of all inner surfaces of the OLCN [5]. Alternatively, synchrotron high-/nano-resolution X-ray computed tomography allows to visualize the OLCN without any staining but is limited to small bone volumes.

▪ Direct Measures of Mechanical Aspects of Bone Material

In contrast to the indirect estimates of mechanical properties of bone material described above, there are methods available which result in a direct measure of bone material properties. These measures are independent of geometry and have the advantage that they can be obtained in a spatially resolved manner which is highly important in a heterogeneous material such as bone.

Scanning Acoustic Microscopy (SAM): The elastic properties of the sectioned bone sample can be visualized with spatial resolution up to 1 μm. SAM is based on generation and focusing of high-frequency acoustic waves to the bone followed by the detection of the echo signals [53]. The inherent contrast (amplitude or difference of time of flight in thin section) is caused by elastic interactions of the incident acoustic waves with the bone material, in particular with changes in lamellar orientation and local mineral content (◘ Fig. 1.8).

Nanoindentation: Direct mechanical measurements can be also obtained from nanoindentation at the atomic force microscope (AFM) [40]. In a controlled deformation experiment, the tip of the AFM (nanosized indenter) is pushed into the bone block or sectioned sample, and the load-displacement curve is recorded, from which the indentation modulus as well as the hardness can be derived. The latter outcomes were shown to generally correlate with local mineral content, despite quite large variation at a specific mineral content, and revealed a high anisotropy depending on the direction of the lamellar structure.

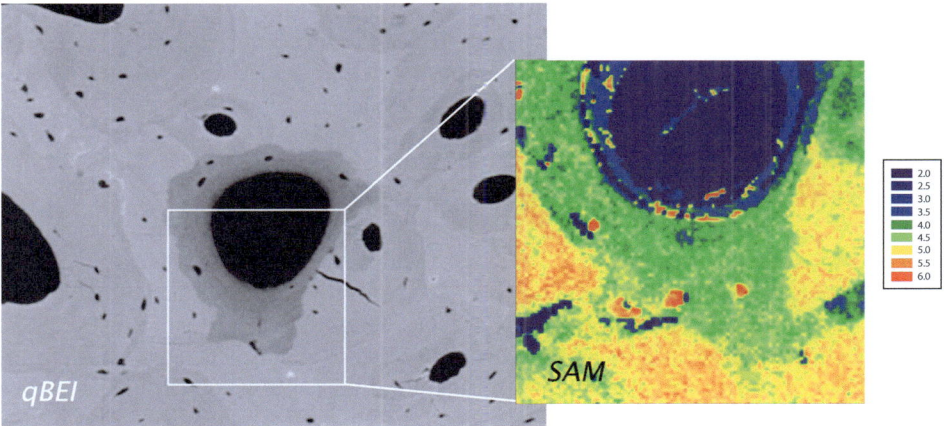

◘ **Fig. 1.8** Backscattered electron mapping of mineral content of osteonal bone (*left*), SAM velocity mapping of enlarged area (*right*). The higher the mineral content, the higher the ultrasound velocities. Both combined informations enable calculation of the local elastic modulus [53]

1.4 Bone Fragility Associated with Altered Bone Material

Commonly, the measurement of bone mineral density (BMD by DXA) is considered for fracture risk assessment and treatment decision for a patient. However, BMD alone is not a perfect predictor of fracture risk. With advancing development of techniques and with awareness of bone material quality being an important contributor to the whole-bone mechanical competence, the characterization/quantification of bone material properties has become an important tool of clinical application (differential diagnosis and treatment monitoring). Two examples are shown in the following.

1.4.1 Bone Material in Postmenopausal Osteoporosis and After Treatment

Postmenopausal osteoporosis is associated with low BMD due to bone loss and reduced bone matrix mineralization (shift of the BMDD to lower calcium concentrations in ◘ Fig. 1.7) as a consequence of increased bone turnover rates [47]. Additionally, altered cross-link patterns have been reported [48]. Commonly, postmenopausal osteoporosis is treated with anti-resorptive agents such as bisphosphonates. Interestingly, relatively small treatment-induced increases in BMD result in disproportionally higher fracture risk reduction of treated patients. The observed increase in degree of mineralization was suggested to be contributing to the improvement of bone quality during treatment [47].

Another strategy in osteoporosis therapy is anabolic treatment, for instance, with teriparatide or parathyroid hormone which are improving BMD by increasing bone volume while even transiently decreasing average bone matrix mineralization (due to large amounts of young BSUs). Another therapy to be mentioned in this context was the previous treatment with sodium fluoride. This treatment was not successful, as it had no positive effect on fracture risk reduction although it caused large increases in bone volume. Analysis of bone material formed under therapy revealed highly increased mineral content and an abnormal nanocomposite including increased size of the mineral particles [31].

1.4.2 Bone Material in Osteogenesis Imperfecta

Osteogenesis imperfecta (OI) is a genetic disease associated with collagen fibril alterations and characterized by increased bone turnover and brittle bone phenotype [49]. The bone fragility is based on reduced tissue quality (reduced bone volume, abnormal architecture) and in particular on deviations from normal bone material quality. Decreased mechanical competence of collagen matrix and hypermineralization (shift of the BMDD to higher calcium concentrations in ◘ Fig. 1.7) has been linked to the increased brittleness in this disease [50]. Bisphosphonate treatment was found beneficial to patients with OI [49, 51]. The positive effect on fracture risk reduction in young patients with OI is based primarily on the increase in cortical width and cancellous bone volume [51] in these patients, while the bone material quality remains unaffected by treatment [52].

Take-Home Messages

- Bone is a hierarchically organized material.
- Each hierarchical level is mechanically optimized and thus important in bone deformation processes.
- One has to distinguish between whole-bone mechanical properties and material properties. The latter are independent of bone size and geometry.
- Bone material quality parameters are usually measured in bone samples in a spatially resolved manner and characterize structure and composition as well as hardness and elasticity of the bone material.
- In addition to lower bone volume/mass altered deformation mechanisms due to altered bone material quality has to be considered in bone fragility.
- Certain bone diseases and treatment thereof alter bone material quality parameters.

References

1. Fratzl P, Weinkamer R. Nature's hierarchical materials. Prog Mater Sci. 2007;52:1263–334.
2. Fratzl P, Gupta HS, Paschalis EP, Roschger P. Structure and mechanical quality of the collagen-mineral nano-composite in bone. J Mater Chem. 2004;14:2115–23.
3. Akkus O, Polyakova-Akkus A, Adar F, Schaffler MB. Aging of microstructural compartments in human compact bone. J Bone Miner Res. 2003;18:1012–9.
4. Fuchs RK, Allen MR, Ruppel ME, Diab T, Phipps RJ, Miller LM, Burr DB. In situ examination of the time-course for secondary mineralization of Haversian bone using synchrotron Fourier transform infrared microspectroscopy. Matrix Biol. 2008;27:34–41.
5. Kerschnitzki M, Wagermaier W, Roschger P, Seto J, Shahar R, Duda GN, Mundlos S, Fratzl P. The organization of the osteocyte network mirrors the extracellular matrix orientation in bone. J Struct Biol. 2011;173(2):303–11.
6. Dong P, Haupert S, Hesse B, Langer M, Goutteno re P, Bousson V, Peyrin F. 3D osteocyte lacunar morphometric properties and distributions in human femoral cortical bone using synchrotron radiation micro-CT images. Bone. 2014;60:172–85.
7. Ebacher V, Guy P, Oxland TR, Wang R. Sub-lamellar microcracking and roles of canaliculi in human cortical bone. Acta Biomater. 2012;8:1093–100.
8. Buenzli PR, Sims NA. Quantifying the osteocyte network in the human skeleton. Bone. 2015;75:144–50.
9. Gupta HS, Seto J, Wagermaier W, Zaslansky P, Boesecke P, Fratzl P. Cooperative deformation of mineral and collagen in bone at the nanoscale. Proc Natl Acad Sci U S A. 2006;103(47):17741–6.
10. Wagermaier W, Gupta HS, Gourrier A, Burghammer M, Roschger P, Fratzl P. Spiral twisting of fiber orientation inside bone lamellae. Biointerphases. 2006;1:1–5.
11. Weiner S, Traub W. Organization of hydroxyapatite crystals within collagen fibrils. FEBS Lett. 1986;206:262–6.
12. Hodge AJ, Petruska JA. Recent studies with the electron microscope on ordered aggregates of the tropocollagen macromolecule. In: Ramachandran GN, editor. Aspects of protein structure. New York: Academic Press; 1963. p. 289–300.
13. Arsenault AL, Grynpas MD. Crystals in calcified cartilage and cortical bone of the rat. Calcif Tissue Int. 1988;43:219–25.
14. Fratzl P, Groschner M, Vogl G, Plenk H, Eschberger J, Fratzl-Zelman N, Koller K, Klaushofer K. Mineral crystals in calcified tissues: a comparative study by SAXS. J Bone Miner Res. 1992;9:1651–5.
15. Landis WJ, Silver FH. Mineral deposition in the extracellular matrices of vertebrate tissues: identification of possible apatite nucleation sites on type I collagen. Cells Tissues Organs. 2009;189:20–4.

16. Boskey AL. Matrix proteins and mineralization: an overview. Connect Tissue Res. 1996;35:357–63.
17. Glimcher MJ. The nature of the mineral component of bone and the mechanism of calcification. Instr Course Lect. 1987;36:49–69.
18. Paschalis EP, Mendelsohn R, Boskey AL. Infrared assessment of bone quality: a review. Clin Orthop Relat Res. 2011;469(8):2170–8.
19. Sharir A, Barak MM, Shahar R. Whole bone mechanics and mechanical testing. Veter J. 2008;177:8–17.
20. Currey JD. The mechanical consequences of variation in the mineral content of bone. J Biomech. 1969;2:1–11.
21. Landis WJ, Librizzi JJ, Dunn MG, Silver FH. A study of the relationship between mineral content and mechanical properties of turkey gastrocnemius tendon. J Bone Miner Res. 1995;10:859–67.
22. Turner CH. Biomechanics of bone: determinants of skeletal fragility and bone quality. Osteoporos Int. 2002;13(2):97–104.
23. Roschger P, Paschalis EP, Fratzl P, Klaushofer K. Bone mineralization density distribution in health and disease. Bone. 2008;42:456–66.
24. Jäger I, Fratzl P. Mineralized collagen fibrils – a mechanical model with a staggered arrangement of mineral particles. Biophys J. 2000;79:1737–46.
25. Misof K, Landis WJ, Klaushofer K, Fratzl P. Collagen from the osteogenesis imperfecta mouse model (oim) shows reduced resistance against tensile stress. J Clin Invest. 1997;100:40–5.
26. Knott L, Bailey AJ. Collagen cross-links in mineralizing tissues: a review of their chemistry, function, and clinical relevance. Bone. 1998;22:181–7.
27. Karim L, Vashishth D. Heterogeneous glycation of cancellous bone and its association with bone quality and fragility. PLoS One. 2012;7:e35047.
28. Saito M, Marumo K. Collagen cross-links as a determinant of bone quality: a possible explanation for bone fragility in aging, osteoporosis, and diabetes mellitus. Osteoporos Int. 2010;21:195–214.
29. Paschalis EP, Shane E, Lyritis G, Skarantavos G, Mendelsohn R, Boskey AL. Bone fragility and collagen cross-links. J Bone Miner Res. 2004;19:2000–4.
30. Garnero P. The contribution of collagen crosslinks to bone strength. Bonekey Rep. 2012;1:182.
31. Fratzl P, Roschger P, Eschberger J, Abendroth B, Klaushofer K. Abnormal bone mineralization after fluoride treatment in osteoporosis: a small-angle x-ray-scattering study. J Bone Miner Res. 1994;9:1541–9.
32. Nalla RK, Kruzic JJ, Ritchie RO. On the origin of the toughness of mineralized tissue: microcracking or crack bridging? Bone. 2004;34:790–8.
33. Peterlik H, Roschger P, Klaushofer K, Fratzl P. From brittle to ductile fracture of bone. Nat Mater. 2006;5:52–5.
34. Fratzl-Zelman N, Valenta A, Roschger P, Nader A, Gelb BD, Fratzl P, Klaushofer K. Decreased bone turnover and deterioration of bone structure in two cases of pycnodysostosis. J Clin Endocrinol Metab. 2004;89:1538–47.
35. Koester KJ, Ager JW 3rd, Ritchie RO. The true toughness of human cortical bone measured with realistically short cracks. Nat Mater. 2008;7(8):672–7.
36. Fratzl P. Bone fracture: when the cracks begin to show. Nat Mater. 2008;7(8):610–2.
37. Yao H, Dao M, Carnelli D, Tai K, Ortiz C. Size-dependent heterogeneity benefits the mechanical performance of bone. J Mech Phys Solids. 2011;59:64–74.
38. Tai K, Dao M, Suresh S, Plazoglu A, Ortiz C. Nanoscale heterogeneity promotes energy dissipation in bone. Nat Mater. 2007;6:454–62.
39. Chapurlat RD, Delmas PD. Bone microdamage: a clinical perspective. Osteoporos Int. 2009;20:1299–308.
40. Misof B, Roschger P, Fratzl P. Imaging mineralized tissues in vertebrates. In: Ducheyne P, Healy KE, Hutmacher DW, Grainger DW, Kirkpatrick CJ, editors. Comprehensive biomaterials, vol. 3. Amsterdam: Elsevier; 2011. p. 407–26.
41. Gamsjaeger S, Mendelsohn R, Boskey A, Gourion-Arsiquaud S, Klaushofer K, Paschalis EP. Vibrational spectroscopic imaging for the evaluation of matrix and mineral chemistry. Curr Osteoporos Rep. 2014;12(4):454–64. doi:10.1007/s11914-014-0238-8.
42. Gamsjaeger S, Mendelsohn R, Klaushofer K, Paschalis EP. Vibrational spectroscopic imaging of hard tissues. In: Salzer R, Siesler HW, editors. Infrared and Raman spectroscopic imaging. Weinheim: Wiley-VCH; 2014.
43. Paschalis EP, Fratzl P, Gamsjaeger S, Hassler N, Brozek W, Eriksen EF, Rauch F, Glorieux FH, Shane E, Dempster D, Cohen A, Recker R, Klaushofer K. Aging versus postmenopausal osteoporosis: bone composition and maturation kinetics at actively-forming trabecular surfaces of female subjects aged 1 to 84 years. J Bone Miner Res. 2016;31:347–57.

44. Montagner F, Kaftandjian V, Farlay D, Brau D, Boivin G, Follet H. Validation of a novel microradiography device for characterization of bone mineralization. J Xray Sci Technol. 2015;23:201–11.
45. Fratzl P. Imaging techniques: extra dimension for bone analysis. Nature. 2015;527:308–9.
46. Roschger A, Hofstaetter JG, Pemmer B, Zoeger N, Wobrauschek P, Falkenberg G, Simon R, Berzlanovich A, Thaler HW, Roschger P, Klaushofer K, Streli C. Differential accumulation of lead and zinc in double-tidemarks of articular cartilage. Osteoarthr Cartil 2013;21:1707–15.
47. Roschger P, Misof B, Paschalis E, Fratzl P, Klaushofer K. Changes in the degree of mineralization with osteoporosis and its treatment. Curr Osteoporos Rep. 2014;12:338–50.
48. Boskey AL. Bone composition: relationship to bone fragility and antiosteoporotic drug effects. Bonekey Rep. 2013;2:447. eCollection 2013.
49. Lindahl K, Langdahl B, Ljunggren Ö, Kindmark A. Treatment of osteogenesis imperfecta in adults. Eur J Endocrinol. 2014;171:R79–90.
50. Fratzl-Zelman N, Misof BM, Klaushofer K, Roschger P. Bone mass and mineralization in osteogenesis imperfecta. Wien Med Wochenschr. 2015;165:271–7. Review.
51. Rauch F, Travers R, Glorieux FH. Pamidronate in children with osteogenesis imperfecta: histomorphometric effects of long-term therapy. J Clin Endocrinol Metab. 2006;91:511–6.
52. Weber M, Roschger P, Fratzl-Zelman N, Schöberl T, Rauch F, Glorieux FH, Fratzl P, Klaushofer K. Pamidronate does not adversely affect bone intrinsic material properties in children with osteogenesis imperfecta. Bone. 2006;39(3):616–22.
53. Blouin S, Puchegger S, Roschger A, Berzlanovich A, Fratzl P, Klaushofer K, Roschger P. Mapping dynamical mechanical properties of osteonal bone by scanning acoustic microscopy in time-of-flight mode. Microsc Microanal. 2014;20:924–36.

Osteoblasts and Osteocytes: Essentials and Methods

Nadja Fratzl-Zelman and Franz Varga

2.1 From Mesenchymal Stem Cells to Osteocytes – 18

2.2 Inverse Relationship Between Osteogenic and Adipogenic Programming – 19

2.3 From Osteoblasts to Osteocytes – 22

2.4 The Osteocyte – 22

2.5 Some Essential Facts About Osteocytes – 23

2.6 Conclusion – 25

2.7 Experimental Systems to Study Osteoblastic Differentiation and Behavior – 25

References – 30

© Springer International Publishing AG 2017
P. Pietschmann (ed.), *Principles of Bone and Joint Research*, Learning Materials in Biosciences,
DOI 10.1007/978-3-319-58955-8_2

What You Will Learn in This Chapter

This chapter discusses the development of the bone-forming cells that are all descendants from the mesenchymal stem cells (MSCs). MSCs have the ability to self-renew and provide a pool for osteoprogenitors. However, MSCs can also differentiate into cells of the mesodermal cell line, which besides the bone-forming cells include chondroblasts, adipocytes, and muscle cells. Hormones, local factors, and the extracellular matrix program the MSCs into the distinct differentiation pathways. Especially, the inverse relationship between osteogenesis and adipogenesis plays a pivotal role for bone formation and maintenance of the bone. During differentiation of the osteoblastic lineage, cells pass distinct states with distinct roles in the bone-forming process, i.e., matrix synthesis and mineralization as well as regulation of bone remodeling which appears to be mainly directed by osteocytes. Moreover, osteocytes have important endocrine functions as they secrete factors into circulation that regulate other organs of the body.

In the second part of this chapter, several experimental systems to study bone cell differentiation and mineralization are presented and discussed.

2.1 From Mesenchymal Stem Cells to Osteocytes

Osteoblasts, bone lining cells, osteocytes, chondrocytes, adipocytes, myoblasts, and fibroblasts differentiate all from common precursor cells present in the non-hematopoietic compartment of the bone marrow, *the mesenchymal stem cells* (MSCs). Cells committed to skeletal cells are also referred to as adventitial reticular cells (ARCs) or CXCL12-abundant reticular (CAR) cells in the murine bone marrow [1]. MSCs reside in the trabecular space (stem cell niche) in common with hematopoietic stem cells, which are the founder cells of the hematopoietic lineage, the source for the bone removing osteoclasts. Additionally, endothelial cells and their progenitors reside in the stem cell niche as well.

Differentiation of the MSCs is regulated on the one hand by a tight interaction between the cells but also by local and hormonal factors activating or repressing gene transcription. This happens by cell surface-bound receptors, which can interact either with cell surface-bound ligands or mobile ligands. Moreover, the receptors might be located intracellularly within the cytoplasm as found with some steroid hormones like the estradiol receptor, moving into the nuclei after hormone binding or directly bound to DNA in the nucleus as the thyroid receptor. In addition, the MSCs are wrapped with extracellular matrix (ECM), which not only stabilizes the three-dimensional structure of the bone marrow but also signals to the cells. Balanced concentrations of hormonal and local factors as well as proper ECM in the trabecular space are therefore important for the development not only for the bone cells but also for hematopoiesis. It is now generally accepted that osteoblasts influence hematopoiesis and vice versa [2, 3] (◘ Fig. 2.1).

Osteoblasts regulate not only differentiation and function of osteoclasts (as discussed below) but also of B-cell differentiation [4] and leukemogenesis [5].

Importantly, MSCs committed to differentiation into skeletal cells are able to differentiate into adipocytes as well. With increasing age there is a bias to adipocyte differentiation, which is manifested in accumulation of adipocytes in the bone marrow [6]. It is

Osteoblasts and Osteocytes: Essentials and Methods

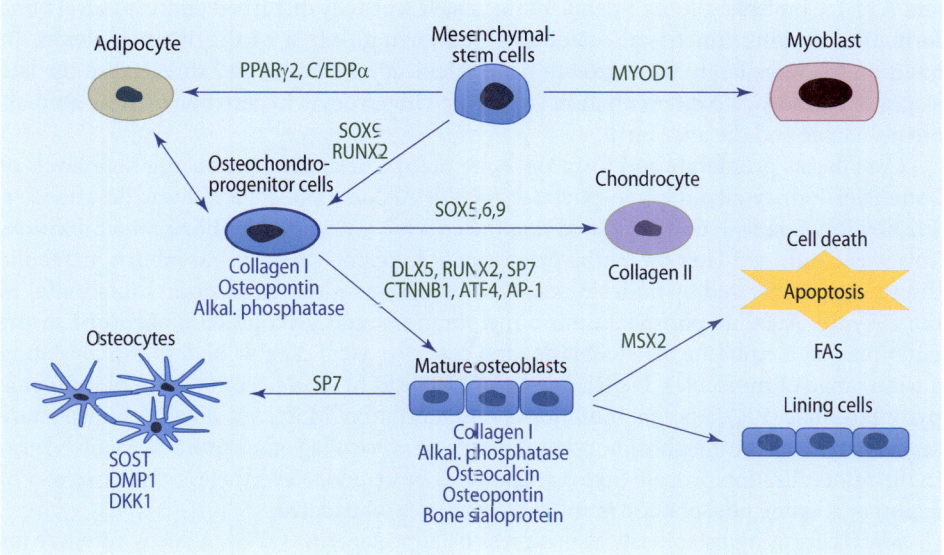

Fig. 2.1 Mesenchymal stem cells have the capacity to differentiate into several cell types (tissues). These processes are regulated by transcription factors (*green*). This results in expression of genes typical or specific for these tissues (*blue*). The surplus of mature osteoblasts, which do not become neither a lining cell nor an osteocyte, is removed by apoptosis

estimated that the bone marrow from newborn infants lacks any fat deposition, while about 70% of the adult bone marrow is occupied by fat. This process is probably irreversible and results in a reduced capacity of bone formation and hematopoiesis [1].

2.2 Inverse Relationship Between Osteogenic and Adipogenic Programming

After commitment the osteoblast precursor cells differentiate along the osteoblastic lineage that involves the development of *osteo-chondroprogenitor* cells representing the common precursor of osteoblast and chondrocytes. These osteo-chondroprogenitors differentiate either into *chondroblasts or osteoblasts*. These cells express two master transcription factors, SOX9 and Runt-related transcription factor 2 (RUNX2) that are essential for chondrogenesis and osteogenesis, respectively, and interact mutually [7–9]. RUNX2 is the master gene of bone formation (although not sufficient for osteoblast maturation) directing MSCs to the osteogenic lineage and inhibiting differentiation into the adipocyte fate. SOX9 is the transcriptional activator of chondrocyte-specific genes. In the osteoprogenitor cells, SOX9 and RUNX2 are co-expressed, whereas SOX2 represses the activity of RUNX2. In proliferating preosteoblastic cells, SOX9 is not any more expressed, and RUNX2 directs the osteogenic cells toward the mature osteoblastic phenotype promoting matrix synthesis and maturation in concert with many other factors like Osterix and ATF4. Conversely, *Runx2*-deficient mice have a heavily disturbed endochondral bone formation, lacking functional osteoblasts and having only a cartilaginous skeleton. In addition the mice fail to form growth plate calcification since Runx2 directs also the late stages of chondrocyte differentiation triggering chondrocyte hypertrophy. These animals are not viable and die after birth.

Osteoblasts proliferate and form the bone matrix. They start producing high levels of bone/liver/kidney alkaline phosphatase *(ALPL)*. Alkaline phosphatase was described in 1923 by Robison who first suggested that the enzyme is essential for bone mineralization. This was confirmed later on by the discovery of hypophosphatasia, an inborn metabolic disorder characterized by undermineralized bone due to loss-of-function mutation(s) of the enzyme. Alkaline phosphatase is a membrane-bound glycoprotein, anchored in the outer plasma membrane of osteoblasts (and chondrocytes) capable of dephosphorylating a wide range of molecules. Deficiencies of ALPL lead to local accumulation of inorganic pyrophosphate (PP_i), a potent inhibitor of mineralization. Moreover, alkaline phosphatase seems necessary for the phosphorylation status of non-collagenous proteins involved also in the mineralization process (like osteopontin; see below). Nevertheless, the exact way of action of alkaline phosphatase is still not completely elucidated.

Matrix-forming osteoblasts also express osteoprotegerin (OPG), a decoy receptor for RANK-ligand that blocks osteoclast formation (for further details, see, e.g., [10]).

In addition, osteoblasts have receptors for many systemic hormones and most importantly for the traditional mineral-regulating hormones (parathyroid hormone (PTH), parathyroid hormone-like hormone, calcitonin, 1,25-dihydroxyvitamin D3, thyroid hormones, growth hormones, androgens, estrogens, etc.).

During extracellular matrix secretion, the cells synthesize collagen and their accessory proteins to form stable collagen fibrils. This includes proteins for post-translational collagen modification (procollagen-lysine, 2-oxoglutarate-5-dioxygenases, PLOD1–PLOD3) important for folding and cross-linking (lysyl oxidase, LOX). Additionally the cells produce non-collagenous proteins (NCPs) with different functions, which are integrated into the bone matrix.

Here are some important NCPs. For more details, refer, for example, to [11]:
- Growth factors like TGFß, bone morphogenetic proteins, and IGF1. These proteins are often integrated as precursors and/or associated with binding proteins and play multiple roles in cell signaling.
- Osteonectin is the most abundant NCP. It is a phosphorylated glycoprotein (35-45 kD protein) regulating collagen organization probably mediating mineral deposition.
- Proteoglycans are macromolecules containing acidic polysaccharide side chains (glycosaminoglycans) attached to a central core protein. In the bone (and skin!), the predominant proteoglycans are decorin, lumican, and biglycan belonging both to the family of secreted small leucine-rich proteoglycans (SLRPs). These proteins bind to collagen and regulate the activity of TGF-β as well as of other growth factors. Perlecan is a very large heparin sulfate proteoglycan with a core protein of over 4000 amino acids and plays an essential role for the maintenance of osteocyte functionality [12].
- Osteopontin, bone sialoprotein, dentin matrix protein 1 (DMP1), and matrix extracellular phosphoglycoprotein (MEPE) belong all to the small integrin-binding ligand N-linked glycoprotein (SILBING) superfamily. They are mostly

highly phosphorylated and play an important role in initiation and regulation of mineralization. Note that DMP1 and MEPE are important regulators of osteocyte function [13].
- Osteocalcin (or bone Gla protein) is a gamma-carboxyglutamic acid-containing 5 kD protein. Osteocalcin is exclusively expressed by mature osteoblasts, binds strongly to hydroxyapatite the mineral of the bone matrix, and is thought to have multiple functions in regulating bone turnover. More recently, it has emerged that osteocalcin is not only stored in the bone matrix but also released into the circulation, acting as a glucose metabolism regulating hormone on pancreatic B-cells to enhance insulin production and secretion and on muscle cells to increase insulin sensitivity and glucose uptake and decreasing visceral fat [14].

It is important to note that metabolically active, matrix-secreting osteoblasts do not function individually but are found in clusters on the bone surface where they deposit new collagen and non-collagenous proteins within a cavity that has been previously resorbed by osteoclasts (see ▶ Chap. 1). Not all matrix-forming osteoblasts present on the bone surface will share the same fate:
- During matrix formation, some cells stop making matrix, are left behind the other active osteoblasts, become embedded within the (still non-mineralized!) organic matrix, and will differentiate toward the *osteocyte* phenotype.
- At the end of the matrix formation process (e.g., when the resorption cavity on the bone surface is filled), alkaline phosphatase activity declines, and some cells will become flatter and metabolically less active. These cells become *lining cells*, form tight junctions with each other, and cover the bone surface forming a natural barrier toward the bone marrow space or stem cell niche. The lining cells, although considered as postosteoblastic, are quiescent cells that retain the ability to redifferentiate into matrix-forming osteoblasts upon exposure hormones or mechanical conditions [15]. As lining cells are connected to the osteocyte via the canalicular network and gap junctions, they could also signal to osteocytes when stress and damage are sensed (see below). Other functions attributed to the lining cells are regulation of the influx and efflux of mineral ions [16] and the ability of cleaning and deposition of a thin layer of a collagenous matrix along the Howship's lacuna to enable new matrix formation [17].
- However, most of the former active osteoblasts will undergo apoptosis and express genes regulating apoptosis. Apoptosis (programmed cell death, very different from necrosis) is a regulated process to maintain bone homeostasis. It is important to realize that the balance of osteoblast proliferation, differentiation, and apoptosis determines the size of a cell population [18]. Conversely, the lifespan of the osteoblasts determines the amount of bone that is formed and can be controlled physiologically by hormones and local factors. For example, intermittent treatment with PTH prevents apoptosis of osteoblasts and osteocytes leading to an increase in bone mass. Similarly, androgens decrease the rate of apoptosis in osteoblasts and osteocytes as other bone anabolic agents like insulin-like growth factors (IGFs) do. Pharmacologic levels of glucocorticoids, however, induce apoptosis of osteoblasts and osteocytes, and this is thought to be the mechanism by which these steroids cause bone loss [19] (◘ Fig. 2.2).

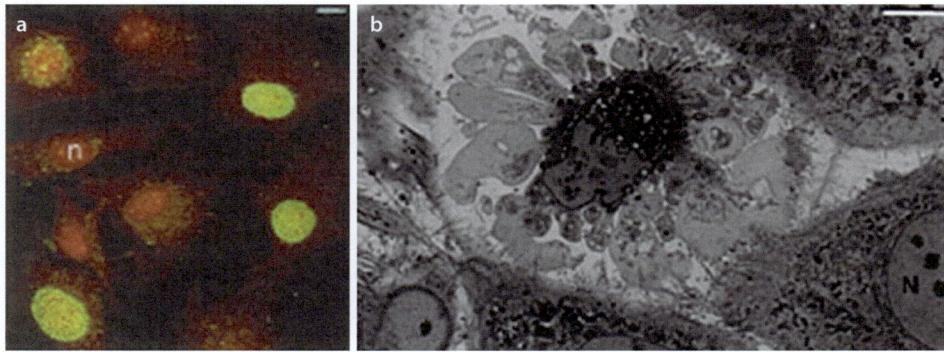

Fig 2.2 Apoptosis is an actively and genetically controlled process, which is ATP dependent. Characteristically, apoptosis leads to enzymatically controlled DNA fragmentation. **a** This fragmentation can be detected by fluorescent TUNEL assay, which uses enzymatic addition of FITC-labeled Bromodeoxyuridine (*green*) to the free 3′-hydroxyl termini of DNA fragments. The cell nuclei are stained with propidium iodide (*red*). **b** Morphologically, apoptotic cells undergo shrinkage and separation from their neighbors; plasma membrane blebbings and a characteristic form of chromatin condensation occur; there is nuclear membrane breakdown and cytolysis into condensed apoptotic bodies which are phagocytized by surrounding cells and macrophages (From Varga et al. [39], with permission from BioScientifica Ltd)

2.3 From Osteoblasts to Osteocytes

The question whether the mature osteoblast-directed mineralization in vitro is physiological or not might be overcome by the assumption that the mineralizing cell in vivo is rather the osteocyte than the osteoblast.

Mikuni-Takagaki et al. [20] characterized already in 1995 the different cell population in newborn rat calvaria after sequential digestion with collagenase and made the following observations:

1. The mature osteoblastic cells on the bone surface do not mineralize but rather separate themselves from the mineralization front by a 10–20 μm layer of unmineralized matrix (8–10 μm in adult remodeled bone).
2. The mineralizing cells are not (or very weakly) positive for alkaline phosphatase.
3. The initiation of mineralization coincides with the phenotypic transformation of cuboidal osteoblastic cells to stellate osteocytes (formation of dendritic processes) within the collagenous matrix, a differentiation state qualified as osteoid-osteocyte.

It has to be stated that about one decade ago, very little was known about osteocyte function. One reason is that unlike osteoblasts, the in vitro study of osteocytes is complicated by the fact that isolated osteocyte from bone tissue does not proliferate [20, 21]. The establishment of osteocyte-like cell lines has greatly improved the knowledge about osteocyte differentiation and function [22].

2.4 The Osteocyte

Osteocytes have been defined during decades by their morphology (cells with cytoplasmic processes) and location (cells embedded within the mineralized bone matrix). They were thought to be passive cells that become «buried alive» in the matrix formed by mature osteoblasts.

Fig. 2.3 Osteocyte network in osteonal equine bone (With courtesy of the Max Planck Institute of Colloids and Interfaces, Department of Biomaterials, Golm, Germany For further details see Kerschnitzki et al. [46])

One of the first suggested functions postulated was that osteocyte senses mechanical deformation. Julius Wolff in 1867 suggested that bone adapts its external shape and internal structure in response to the mechanical forces that are required to support it. The remodeling of the bone in response to loading is achieved via mechano-transduction, a process through which forces or other mechanical signals are converted into biochemical signals (Fig. 2.3).

2.5 Some Essential Facts About Osteocytes

- They are differentiated cells of the osteoblastic lineage that become embedded within the mineralizing matrix.
- They share many markers with osteoblasts but do not make matrix.
- They are the most abundant (90–95%) and the longest-lived cells in the bone. Their number in the human adult skeleton is estimated to 42 billion [23].
- They are connected through dendritic processes called canaliculi (about 89 ± 25 per cell, in total human skeleton about 3.7 trillion! [23]) via gap junctions (= transmembrane channels that connect cytoplasm of two adjacent cells allowing the passage of molecules <1 kDa) [24]:
  - To each other
  - To cells on the bone surface (osteoblasts, lining cells)
  - To the bone marrow (osteoblast and osteoclast recruitment!)
  - To blood vessels (!)

- Cell body and dendritic processes form a functional network, the lacunocanalicular system, which is surrounded by an unmineralized space filled with interstitial fluid providing oxygen and nutrients to maintain cell viability in the mineralized environment. Dendritic osteocytes convert from polygonal matrix-producing mature osteoblasts by progressing through different transitional stages and sequential expression of marker genes reflecting changes in morphology and functionality.

Osteocyte differentiation stage	Some important marker genes	Function
Young osteoid-osteocyte: Cell embedded in non-mineralized matrix, beginning to generate dendritic processes	Podoplanin (*PDPN*): transmembrane protein, the earliest known marker of osteocyte differentiation	Regulates formation of dendritic processes
	Membrane-anchored proteinase that cleaves collagen (*MMP14*)	Important for dendritic formation and morphology
Late mineralizing osteoid-osteocyte: Cell embedded within the osteoid with small, calcified spheres forming along the cell membrane toward the mineralization front	Dentin matrix protein1 (*DMP1*): secreted serin-rich acidic protein with many phosphorylation sites	Regulates osteocyte maturation, phosphate metabolism, and mineralization Inactivation mutations cause autosomal recessive hypophosphatemia and osteomalacia
	Phosphate-regulating gene with homologies to endopeptidase on the X chromosome (*PHEX*)	Regulates osteocyte maturation, phosphate homeostasis, and mineralization Inactivation mutations cause hypophosphatemic rickets (X-linked hypophosphatemia (XLH))
	Matrix extracellular phosphoglycoprotein (*MEPE*)	Regulates phosphate metabolism and mineralization. Inhibits PHEX
	Osteopontin (*SPP1*)	Negative regulator of bone mineralization
Mature osteocyte: Cell is completely embedded within the mineralized matrix Numerous dendritic processes connect the osteocytes to each other	Sclerostin, secreted factor. Highly and specifically expressed in the late osteocyte (*SOST*)	Negative regulator of bone formation through inhibition of the WNT/ß-catenin signaling pathway (regulation of transcription factors) in a negative feedback loop [25]. Treatment of mice with a sclerostin antibody leads to increased osteoblast number by converting quiescent lining cells into active osteoblasts [26] Inactivation mutations cause sclerostosis and van Buchem disease with increased bone formation

Osteocyte differentiation stage	Some important marker genes	Function
	Fibroblast growth factor 23 (*FGF23*) Secreted factor! Most important organ is the kidney Highly expressed in *DMP1* and *PHEX*-associated hypophosphatemic rickets, chronic kidney disease, and tumor-induced osteomalacia	Reduces serum phosphate (Pi) levels by inhibiting renal phosphate reabsorption and downregulation of 1,25-dihydroxyvitamin D3 synthesis *The osteocyte network becomes an endocrine gland* [14] Inactivation mutation causes autosomal dominant hypophosphatemic rickets
	Receptor activator of nuclear factor – κB ligand (RANKL, *TNFSF11*)	Control of osteoclastogenesis – major contribution to bone remodeling in adults [27]
	Tartrate-resistant acid phosphatase (TRAP, *ACP5*) Cathepsin K (*CTSK*)	Removal of perilacunar matrix = osteocyte osteolysis. Important in situation of high calcium demand like lactation [27]
	hypoxia upregulated 1 (ORP150, *HYOU1*)	Preserves viability of osteocytes in hypoxic environment

2.6 Conclusion

Emerging data from osteocyte function have established a new paradigm: osteocytes embedded within the mineralized bone matrix are extremely active and multifunctional cells – they control bone mineralization mainly through expression of factors that regulate phosphate homeostasis (reviewed by [22]); they secrete factors that target the kidney and muscles; they do remodel their extracellular matrix and modify their microenvironment. Osteocytes regulate bone remodeling through regulation of both osteoclasts and osteoblasts as well as their apoptosis, which is an essential tool to control skeletal damage repair [28]. Most importantly, the lacunocanalicular system appears as a highly organized network of connected osteocytes that sense mechanical strain, respond to chemical signals, and orchestrate bone homeostasis. The discovery of these multiple signaling pathways raises also the possibility to develop new therapeutic agents for skeletal diseases [29].

2.7 Experimental Systems to Study Osteoblastic Differentiation and Behavior

In their pioneering work, Friedenstein and coworkers showed for the first time that a bone marrow cell suspension contains a subset of long, plastic-adherent cells with a robust proliferative activity that will give rise to single cell-derived colonies (or colony-forming units, CFUs) with the capacity to differentiate into the bone, chondrocytes, adipocytes, and fibroblastic cells [6, 30, 31]. Experimentally, these cells can be used to study osteoblastic differentiation and regulation. Cell behavior depends on their (micro)environment: the substrate, the degree of contact with other cells, the constitution of the medium, the oxygen tension, and more. Under optimal culture, cells proliferate and differentiate in vitro to form an extracellular matrix that might become later on mineralized (reviewed by, i.e., [32]).

Actually, primary bone cells or cell lines from chick, rat, and mouse are widely used to study the molecular properties of the osteoblast phenotype during proliferation, differentiation, and maturation. Osteoblasts and early osteocytes can be isolated from aseptically dissected calvaria or long bones. For this purpose the bones are serially digested with collagenase. After each sequential digestion, the cell suspension is precipitated by centrifugation and, after washing the cells, suspended in culture medium and seeded into cell culture dishes; the last fraction shows the phenotype of early osteocytes.

Another experimental system is a culture of cells growing out from trabecular bone. For this purpose, small pieces of trabecular bone, i.e., remnants from surgery, are placed in culture dishes containing an appropriate medium for about 3–4 weeks. Thereafter, a lawn of cells is found that can be split several times, and these cells can be frozen in liquid nitrogen for long-term storage.

Osteosarcoma cell lines with diverse differentiation stages can be used as well for studying osteoblastic cell behavior. It must be emphasized that these osteoblast-like cell lines are established from tumor and consequently the system might have some limitations. In particular, the genotype of these transformed and immortalized cell lines is often partially polyploid, and some genes are possibly overrepresented, whereas some others may be deleted or mutated. Nevertheless, these cells are very suitable for studying specific issues. The most frequently used cell lines are U-2 OS, MG-63, and SaOS-2. They differentially respond to vitamin D. While MG-63 expresses and responds to vitamin D with increase expression and secretion of osteocalcin, the marker protein of the mature osteoblast and the two other cell lines do not. Mineralization, however, was demonstrated for MG-63 and SaOS-2 [33].

For mice, several cell lines exist showing different phenotypes and differentiation states and competence. The cell line with the curious denomination C3H10T1/C3H10T2 is an undifferentiated mesenchymal cell line with the potential to differentiate into myoblasts [34], adipocytes [35], chondrocytes, and osteoblasts [36]. This cell line is best suited to study differentiation between osteoblasts and adipocytes [35]. A more differentiated cell line on the way to osteoblasts is the stromal cell line ST2. This cell line, as suggested for stromal cells, supports the differentiation of osteoclast-like cells [37].

An interesting and widely used cell line is MC3T3-E1 and will be discussed here a little more in detail: the cell line was established by immortalization of newborn mouse osteoblasts and behaves similar as primary cells isolated from newborn mouse calvaria by sequential digestion with collagenase [38]. The MC3T3-E1 cells have the capacity to differentiate into mature osteoblasts as indicated by increasing expression levels of alkaline phosphatase and osteocalcin [39]. During prolonged culture time (up to 6 weeks), the cells form many cell layers with a tissue-like appearance [40, 41]. Moreover, the cells form discrete three-dimensional nodular structures consisting of collagenous matrix and covered with cuboidal alkaline phosphatase positive cells [32, 38]. It is generally assumed that the nodules become mineralized by mature osteoblasts [32].

However, MC3T3-E1 like most osteoblastic cell cultures usually do not mineralize spontaneously their extracellular matrix but generally require phosphate supplementation, to induce mineral deposition. For this purpose, 5–10 mM β-glycerophosphate (βGP) is added (generally designated as mineralizing medium) [42].

Although the process of mineralization is still controversial and yet not clarified, there are some aspects that are general consents. The mineral in the bone is hydroxyapatite, a calcium phosphate compound. The solubility product of Ca^{++} ions with phosphate ions (P_i) of the intercellular liquid favors calcium phosphate precipitation. However, mineralization is an ordered and highly regulated process where delivery of Ca^{++} and phosphate ions occurs only at the mineralization front. Therefore uncontrolled initiation of mineralization must rather be inhibited than promoted. Matrix proteins like osteopontin, SIBLING proteins, as well as enzymes regulating the local ratio between pyrophosphate (PPi inhibits mineralization) and phosphate ions (Pi promotes mineralization) are important players that in concert modulate the physiological mineralization process [43] (◘ Fig. 2.4).

Addition of βGP to MC3T3-E1 cultures provides an increase in Pi concentration, possibly by the action of enzymes synthesized by the cells, like alkaline phosphatase, promoting formation of calcium phosphate. Whether the formed mineral is hydroxyapatite or non-apatitic and ectopic (non-collagen associated) is still a matter of controversies [42] (◘ Fig. 2.5).

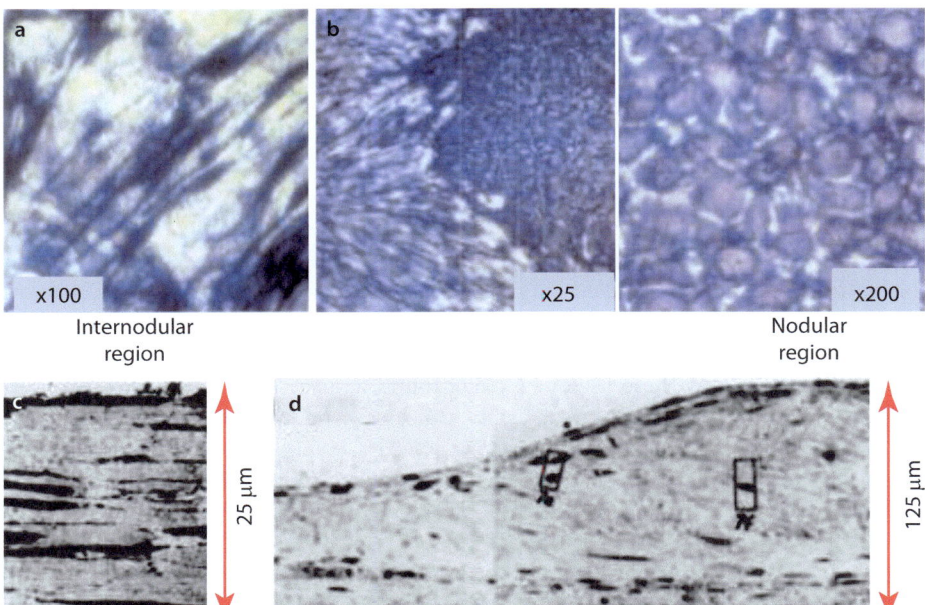

◘ **Fig. 2.4** Long-term culture for 6 weeks of MC3T3-E1 cells in the presence of ascorbate to enable collagen synthesis, which is important for osteoblastic differentiation. Alkaline phosphatase-positive cells are stained in blue. In the internodular region **a**, the predominant phenotype of the cells is a spindle-shaped fibroblastic like phenotype, while in nodular region colonies densely packed cuboidal alkaline phosphatase positive cells were observed **b**, **c** the hallmark of the mature osteoblast. Cross sections of the internodular region **c** and the nodular region **d** (*right* From Fratzl-Zelman et al. [41] with permission from Elsevier)

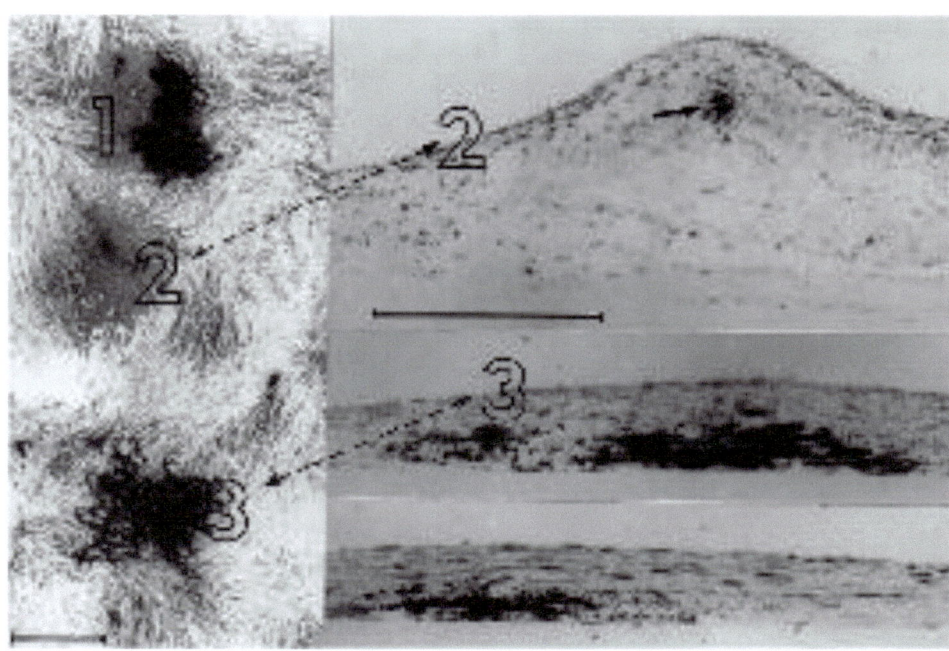

Fig. 2.5 Six-week-old cultures of MC3T3-E1 treated with BGP for the last 2 weeks. The culture was stained with the azo-dye to localize ALPL-positive cells (*gray*) and with von-Kossa staining for simultaneous visualization of mineral deposition (*black*). *1* Nodule with high mineral content. *2* Nodule with low mineral content. *3* Internodular region with mineral deposition. Left overview, scale bar = 1 mm right: cross section, scale bar = 0.2 mm (From Fratzl-Zelman et al. [42] with permission from Elsevier)

Osteocytic cells are highly differentiated cells embedded within the bone matrix and therefore more difficult to isolate and to maintain in culture. In principle, they can be isolated from aseptically dissected long bones of 3- or 4-day-old mice as well. Therefore, bones are serially digested with collagenase, and the isolated cells are recovered by centrifugation and seeded on dishes (see above); the last fraction shows the phenotype of early osteocytes. In fact these osteocytes were found to adhere rapidly to glass or plastic substrates, forming numerous cytoplasmic processes and making contact with other osteocytes [20, 21]. However, these postmitotic cells become rapidly overgrown by the mitotic fibroblast-like cells that were co-extracted from the bone, making further functional analyses very difficult. But there are now recent reports on successful isolation and culture of human osteocytes [44, 45].

Studying osteocytes in culture became much easier with the establishment of osteocyte-like cell lines. Actually, three characteristic osteocyte-like cell lines derived from mouse long bones are available: the post-osteoblast, preosteocyte-like cells MLO-A5 that spontaneously mineralize in culture, the MLO-Y4 cells that mimic a rather mature osteocyte phenotype, and more recently the IDG-SW3 cell line that expresses characteristic markers from late osteoblast to mature osteocyte phenotype in vitro. The study of these cell lines provided within the last few years a wealth of data showing that osteocytes are

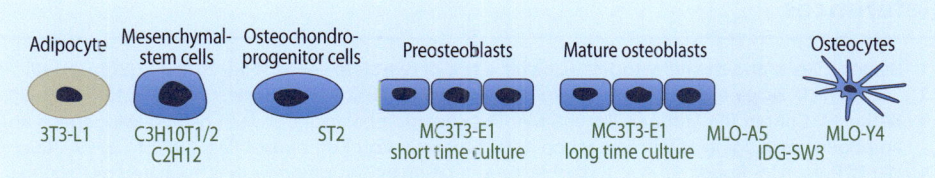

Fig. 2.6 Nearly every phenotype of the osteoblastic differentiation cascade can be approached by an established cell line

extremely active cells with a complex developmental biology and multiple functions in bone metabolism [22].

Taken together, the accurate knowledge about osteoblast phenotype and differentiation shows that the choice of an appropriate cell line is crucial for a correct experimental setup (Fig. 2.6).

> **Take-Home Message**
> - Osteoblasts derive from mesenchymal stem cells of the bone marrow, differentiate in a highly regulated fashion, and secrete an extracellular matrix consisting mainly of type I collagen and a small amount of non-collagenous proteins. Active, matrix-secreting osteoblasts undergo one of the three fates: the great majority dies by apoptosis, which is an important regulatory process to maintain bone homeostasis, and some remain quiescent on bone surfaces as flat lining cells, while about 5–20% differentiate into osteocytes.
> - Osteocytes are the most abundant cells of the bone tissue. As they become embedded within the bone matrix, they undergo a characteristic transformation from a cuboidal to a stellate cell through the formation of multiple dendritic processes called canaliculi and start to mineralize the surrounding matrix. Canaliculi and osteocytic cell body (also called osteocyte lacunae) form a highly functional network, which allows communication among osteocytes, with osteoblasts and lining cells.
> - Osteocytes secrete factors that target also osteoclasts and hormones that affect other organs by endocrine mechanisms. The many functions of osteocytes include mechanosensing, regulation of phosphate metabolism, bone formation, and bone resorption.
> - Cell cultures of osteoblasts and osteocytes allow to determine biological functions; however, one has to be aware that all in vitro models have strengths and limitations.

Acknowledgments The authors are very grateful to Univ. Prof. Dr. Klaus Klaushofer, Director of the Ludwig Boltzmann Institute of Osteology, for the continuous support and many discussions about bone biology and its relevance for the clinic.

The work was supported by the AUVA (Research funds of the Austrian workers compensation board) and by the WGKK (Viennese sickness insurance funds).

References

1. Bianco P. Bone and the hematopoietic niche: a tale of two stem cells. Blood. 2011;117(20):5281–8.
2. Scadden DT. Nice neighborhood: emerging concepts of the stem cell niche. Cell. 2014;157(1):41–50.
3. Bethel M, Chitteti BR, Srour EF, Kacena MA. The changing balance between osteoblastogenesis and adipogenesis in aging and its impact on hematopoiesis. Curr Osteoporos Rep. 2013;11(2):99–106.
4. Wu JY, Purton LE, Rodda SJ, et al. Osteoblastic regulation of B lymphopoiesis is mediated by Gs{alpha}-dependent signaling pathways. Proc Natl Acad Sci U S A. 2008;105(44):16976–81.
5. Kode A, Manavalan JS, Mosialou I, et al. Leukaemogenesis induced by an activating beta-catenin mutation in osteoblasts. Nature. 2014;506(7487):240–4.
6. Chen Q, Shou P, Zheng C, et al. Fate decision of mesenchymal stem cells: adipocytes or osteoblasts? Cell Death Differ. 2016;23(7):1128–39.
7. Karsenty G. Transcriptional control of skeletogenesis. Annu Rev Genomics Hum Genet. 2008;9:183–96.
8. Akiyama H, Kim JE, Nakashima K, et al. Osteo-chondroprogenitor cells are derived from Sox9 expressing precursors. Proc Natl Acad Sci U S A. 2005;102(41):14665–70.
9. Zhou G, Zheng Q, Engin F, et al. Dominance of SOX9 function over RUNX2 during skeletogenesis. Proc Natl Acad Sci U S A. 2006;103(50):19004–9.
10. Khosla S. Minireview: the OPG/RANKL/RANK system. Endocrinology. 2001;142(12):5050–5.
11. Sroga GE, Vashishth D. Effects of bone matrix proteins on fracture and fragility in osteoporosis. Curr Osteoporos Rep. 2012;10(2):141–50.
12. Thompson WR, Modla S, Grindel BJ, et al. Perlecan/Hspg2 deficiency alters the pericellular space of the lacunocanalicular system surrounding osteocytic processes in cortical bone. J Bone Miner Res. 2011;26(3):618–29.
13. Staines KA, MacRae VE, Farquharson C. The importance of the SIBLING family of proteins on skeletal mineralisation and bone remodelling. J Endocrinol. 2012;214(3):241–55.
14. Fukumoto S, Martin TJ. Bone as an endocrine organ. Trends Endocrinol Metab. 2009;20(5):230–6.
15. Kim SW, Pajevic PD, Selig M, et al. Intermittent parathyroid hormone administration converts quiescent lining cells to active osteoblasts. J Bone Miner Res. 2012;27(10):2075–84.
16. Neve A, Corrado A, Cantatore FP. Osteoblast physiology in normal and pathological conditions. Cell Tissue Res. 2011;343(2):289–302.
17. Everts V, Delaisse JM, Korper W, et al. The bone lining cell: its role in cleaning Howship's lacunae and initiating bone formation. J Bone Miner Res. 2002;17(1):77–90.
18. Varga F, Luegmayr E, Fratzl-Zelman N, et al. Tri-iodothyronine inhibits multilayer formation of the osteoblastic cell line, MC3T3-E1, by promoting apoptosis. J Endocrinol. 1999;160(1):57–65.
19. Bodine PV. Wnt signaling control of bone cell apoptosis. Cell Res. 2008;18(2):248–53.
20. Mikuni-Takagaki Y, Kakai Y, Satoyoshi M, et al. Matrix mineralization and the differentiation of osteocyte-like cells in culture. J Bone Miner Res. 1995;10(2):231–42.
21. van der Plas A, Nijweide PJ. Isolation and purification of osteocytes. J Bone Miner Res. 1992;7(4):389–96.
22. Dallas SL, Prideaux M, Bonewald LF. The osteocyte: an endocrine cell ... and more. Endocr Rev. 2013;34(5):658–90.
23. Buenzli PR, Sims NA. Quantifying the osteocyte network in the human skeleton. Bone. 2015;75:144–50.
24. Batra N, Kar R, Jiang JX. Gap junctions and hemichannels in signal transmission, function and development of bone. Biochim Biophys Acta. 2012;1818(8):1909–18.
25. Baron R, Kneissel M. WNT signaling in bone homeostasis and disease: from human mutations to treatments. Nat Med. 2013;19(2):179–92.
26. Kim SW, Lu Y, Williams EA, et al. Sclerostin antibody administration converts bone lining cells into active osteoblasts. J Bone Miner Res. 2016;32(5):892–901.
27. Xiong J, O'Brien CA. Osteocyte RANKL: new insights into the control of bone remodeling. J Bone Miner Res. 2012;27(3):499–505.
28. Jilka RL, Noble B, Weinstein RS. Osteocyte apoptosis. Bone. 2013;54(2):264–71.
29. Plotkin LI, Bellido T. Osteocytic signalling pathways as therapeutic targets for bone fragility. Nat Rev Endocrinol. 2016;12(10):593–605.
30. Owen M, Friedenstein AJ. Stromal stem cells: marrow-derived osteogenic precursors. Ciba Found Symp. 1988;136:42–60.
31. Bianco P, Robey PG, Simmons PJ. Mesenchymal stem cells: revisiting history, concepts, and assays. Cell Stem Cell. 2008;2(4):313–9.
32. Aubin JE. Regulation of osteoblast formation and function. Rev Endocr Metab Disord. 2001;2(1):81–94.

33. Fedde KN. Human osteosarcoma cells spontaneously release matrix-vesicle-like structures with the capacity to mineralize. Bone Miner. 1992;17(2):145–51.
34. Davis RL, Weintraub H, Lassar AB. Expression of a single transfected cDNA converts fibroblasts to myoblasts. Cell. 1987;51(6):987–1000.
35. Ahrens M, Ankenbauer T, Schroder D, Hollnagel A, Mayer H, Gross G. Expression of human bone morphogenetic proteins-2 or −4 in murine mesenchymal progenitor C3H10T1/2 cells induces differentiation into distinct mesenchymal cell lineages. DNA Cell Biol. 1993;12(10):871–80.
36. Wang EA, Israel DI, Kelly S, Luxenberg DP. Bone morphogenetic protein-2 causes commitment and differentiation in C3H10T1/2 and 3T3 cells. Growth Factors. 1993;9(1):57–71.
37. Udagawa N, Takahashi N, Akatsu T, et al. The bone marrow-derived stromal cell lines MC3T3-G2/PA6 and ST2 support osteoclast-like cell differentiation in cocultures with mouse spleen cells. Endocrinology. 1989;125(4):1805–13.
38. Sudo H, Kodama HA, Amagai Y, Yamamoto S, Kasa S. In vitro differentiation and calcification in a new clonal osteogenic cell line derived from newborn mouse calvaria. J Cell Biol. 1983;96(1):191–8.
39. Varga F, Rumpler M, Luegmayr E, Fratzl-Zelman N, Glantschnig H, Klaushofer K. Triiodothyronine, a regulator of osteoblastic differentiation: depression of histone H4, attenuation of c-fos/c-jun, and induction of osteocalcin expression. Calcif Tissue Int. 1997;61(5):404–11.
40. Luegmayr E, Varga F, Frank T, Roschger P, Klaushofer K. Effects of triiodothyronine on morphology, growth behavior, and the actin cytoskeleton in mouse osteoblastic cells (MC3T3-E1). Bone. 1996;18(6):591–9.
41. Fratzl-Zelman N, Horandner H, Luegmayr E, et al. Effects of triiodothyronine on the morphology of cells and matrix, the localization of alkaline phosphatase, and the frequency of apoptosis in long-term cultures of MC3T3-E1 cells. Bone. 1997;20(3):225–36.
42. Fratzl-Zelman N, Fratzl P, Horandner H, et al. Matrix mineralization in MC3T3-E1 cell cultures initiated by beta-glycerophosphate pulse. Bone. 1998;23(5):511–20.
43. Zhou X, Cui Y, Han J. Phosphate/pyrophosphate and MV-related proteins in mineralisation: discoveries from mouse models. Int J Biol Sci. 2012;8(6):778–90.
44. Prideaux M, Schutz C, Wijenayaka AR, et al. Isolation of osteocytes from human trabecular bone. Bone. 2016;88:64–72.
45. Shah KM, Stern MM, Stern AR, Pathak JL, Bravenboer N, Bakker AD. Osteocyte isolation and culture methods. Bonekey Rep. 2016;5:838.
46. Kerschnitzki M, et al. The organization of the osteocyte network mirrors the extracellular matrix orientation in bone. J Struct Biol. 2011;173:303–11.

Osteoclasts: Essentials and Methods

Nadia Rucci and Anna Teti

3.1 **Osteoclast Biology – 34**
3.1.1 Osteoclastogenesis – 34
3.1.2 Osteoclast Functions – 36

3.2 **Osteoclast Deregulation and Related Pathologies – 39**
3.2.1 Osteoporosis – 39
3.2.2 Bone Metastases – 40
3.2.3 Osteopetrosis – 41

3.3 **Osteoclast Pharmacology – 42**
3.3.1 Bisphosphonates – 43
3.3.2 Denosumab – 43
3.3.3 New Antiresorptive Agents – 44

3.4 **Methods for Osteoclast Primary Cultures – 44**
3.4.1 Osteoclast Isolation from Mouse Bone Marrow Cells – 45
3.4.2 Osteoclast Isolation from Human Peripheral Blood – 45
3.4.3 Evaluation of Osteoclast Differentiation – 46
3.4.4 Evaluation of Osteoclast Function – 47

3.5 **Conclusions – 47**

References – 48

© Springer International Publishing AG 2017
P. Pietschmann (ed.), *Principles of Bone and Joint Research*, Learning Materials in Biosciences,
DOI 10.1007/978-3-319-58955-8_3

What You Will Learn in This Chapter

Osteoclasts are monocyte/macrophage arising cells with the classical function of bone resorption, thus fulfilling the bone remodelling process in concert with osteoblasts. The correct balance between osteogenic functions and osteoclast activity is mandatory to prevent skeletal diseases. While an exacerbated bone resorption is associated with bone loss, eventually leading to osteoporosis, the lack of osteoclast activity is responsible for osteopetrosis, a rare genetic disorder characterised by increased bone density and a wide heterogeneity in terms of severity, ranging from asymptomatic to fatal in infancy. Besides this well-established role in bone resorption, new functions have been recently attributed to the osteoclast. Indeed there is a reciprocal crosstalk between osteoclasts and osteoblasts which influence each other, in case of osteoclasts by releasing factors from the resorbing matrix and by secreting the so-called clastokines. Another recently discovered function of osteoclasts is haematopoiesis regulation. This draws to the obvious consequence that any osteoclast dysfunction would not cause exclusively a bone phenotype. As for other cell types, the knowledge of osteoclast biology has benefited from the study of skeletal diseases in which their formation and function are compromised. Furthermore, well-established methods are available to perform osteoclast primary cultures, and the identification of macrophage colony-stimulating factor (M-CSF) and receptor activator of nuclear factor κB ligand (RANKL) as pro-osteoclastogenic factors fostered their employment. Therefore, nowadays the preferential way to obtain purified differentiated osteoclasts is to isolate osteoclast precursors from the bone marrow or peripheral blood mononuclear cells and treat with the above-mentioned pro-osteoclastogenic cytokines.

3.1 Osteoclast Biology

Osteoclasts are classically described as the cells of the bone tissue devoted to destroy the mineralised matrix, thus accomplishing an apparently damaging function that actually is crucial for the correct homeostasis of this hard tissue [1]. In fact, bone resorption is a necessary step that, when perfectly balanced by the osteogenic function of osteoblasts, fulfils the bone remodelling process, which ensures the maintenance of a correct bone mass throughout the life of each individual, in terms of both quality and quantity. When this delicate osteoblast-osteoclast coupling is unbalanced, it causes bone diseases.

As for many other cell types of the body, the biology of osteoclasts has benefited from the study of skeletal diseases in which their formation and function are compromised. The classical example is osteopetrosis (also known as marble bone disease), a rare genetic disorder characterised by increased bone density accompanied by a wide range of complications, such as bone marrow failure, compressive neuropathies, hypocalcaemia and fractures, most of them resulting from the lack of bone resorption [2].

3.1.1 Osteoclastogenesis

One of the first experiments suggesting the actual origin of osteoclasts dates back to the late 1970s, when it was shown that bone resorption could be restored in osteopetrotic mice by bone marrow transplantation or by parabiosis, thus suggesting a haematopoietic origin and a circulating ability of osteoclast precursors [3]. These results came after other studies that, in contrast, had hypothesised a common origin for osteoblasts and osteoclasts [4].

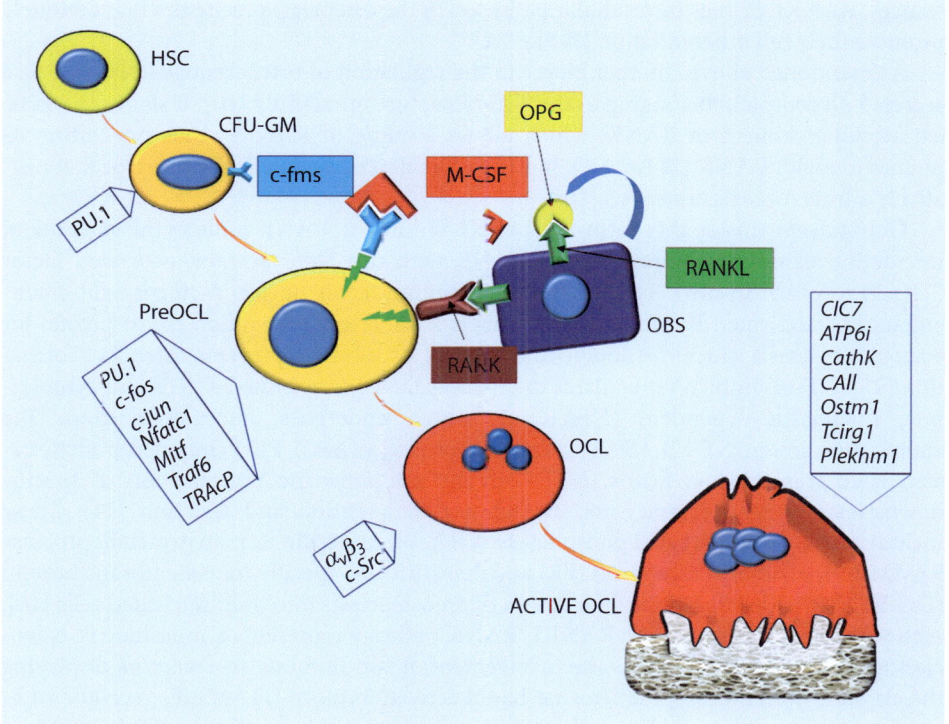

Fig. 3.1 Osteoclast differentiation. Schematic representation of the multistep process of osteoclast differentiation. For each step, the genes crucial for the process are reported (*boxes*). *HSC* haematopoietic stem cells, *GM-CFU* granulocyte/monocyte colony-forming unit, *OCL* osteoclast, *M-CSF* macrophage colony-stimulating factor, *OPG* osteoprotegerin, *RANK* receptor activator of nuclear factor κB, *RANKL* RANK ligand

Therefore, as stated by Chambers in his recent review [5], «the osteoclasts are not really bone cells, but blood-borne immigrants into bone».

By now, it is known that osteoclasts arise from the monocyte/macrophage cell line through a step-by-step process requiring the sequential activation of specific pathways (Fig. 3.1). First, there is the switch-on of the transcription factor PU.1 [6], which drives the positive regulation of the receptor of the macrophage colony-stimulating factor (M-CSF), c-fms, expressed by the haematopoietic stem cells (HSCs). This event eventually leads to cell commitment towards a common progenitor for macrophage and osteoclasts, belonging to the granulocyte macrophage colony-forming unit (CFU-GM) lineage [7, 8]. M-CSF is necessary for proliferation and survival of these macrophage/osteoclast progenitors. Moreover, it promotes the expression of the receptor activator of nuclear factor κB (RANK) [9]. This is a very crucial step due to the fundamental role of the RANK/RANKL/osteoprotegerin (OPG) triad in osteoclastogenesis [10–13]. Moreover, the appearance of RANK also marks the transition from the CFU-GM cells to a committed osteoclast precursor [14]. The pro-osteoclastogenic ligand of RANK (RANKL) is produced by lymphocytes, stromal cells, osteoblasts and osteocytes, preferentially as a transmembrane cytokine, requiring a cell-cell contact, and, in lower quantities, as a soluble factor released through the proteolytic cleavage of the active ectodomain. In both circumstances, RANKL primes intracellular pathways in the osteoclast precursors that definitely promote their full differentiation (◘ Fig. 3.1).

As mentioned above, another player in the regulation of osteoclastogenesis is OPG, a secreted glycoprotein belonging to the TNF receptor superfamily [10]. It shares the same extracellular domain of RANK, which allows binding of RANKL, thus preventing its interaction with RANK (◘ Fig. 3.1). It is therefore described as a decoy receptor that negatively affects osteoclastogenesis [10].

Going deeply inside this pathway, RANKL binding to RANK induces the subsequent interaction of the cytoplasmic tail of RANK with the TNF receptor-associated factor (TRAF) 6, which in turn activates the transcription factor nuclear factor κ-light-chain-enhancer of activated B cells (NF-κB). This is a dimeric transcription factor pivotal for osteoclastogenesis, since the double knockout of its subunits prevents osteoclast formation [15]. NF-κB in turn upregulates the nuclear factor of activated T cells and cytoplasmic, calcineurin-dependent (NFATc) 1, which undergoes auto-amplification. The cooperation among NF-kB, NFATc1, activator protein 1 (AP1), PU.1 and microphthalmia-associated transcription factor (MITF) finally promotes the transcription of specific downstream genes necessary for osteoclast differentiation and function [14]. These include tartrate-resistant acid phosphatase (TRAcP), cathepsin K, matrix metalloprotease 9 (MMP-9), calcitonin receptor (CTR) and dendritic cell-specific transmembrane protein (DC-STAMP), the latter pivotal for fusion of preosteoclasts into multinucleated cells [16]. Although the discovery of the RANKL/RANK pathway represents a milestone in osteoclastogenesis, this also requires the involvement of two immune co-receptors displaying the classical immunoreceptor tyrosine-based activation motif (ITAM): Fc receptor common γ signalling chain (FcRγ) and DNAX-activating protein of 12 kDa (DAP12). These co-receptors interact with osteoclast-associated receptor (OSCAR) and trigger receptor expressed on myeloid cells 2 (TREM2), with a resulting activation of the phospholipase Cγ (PLCγ) eventually leading to intracellular Ca^{2+} oscillations that mediate calcineurin-dependent activation of NFATc1 [17].

3.1.2 Osteoclast Functions

Apart from the well-known activity of bone resorption, the picture of osteoclast duties has changed over the years, thus delineating a new profile including unexpected functions for this very versatile cell, as we will describe in the next paragraphs.

Osteoclast Bone Resorption

The machinery of bone resorption is now well known and requires mature and polarised multinucleated cells, firmly adhering to the bone surface in order to isolate the underlying matrix that will be digested. A mature osteoclast is a polarised cell with plasma membrane domains associated with specific functions. The specialised domain facing the bone matrix, characterised by extensive folding of the plasma membrane, is named «ruffled border». The «sealing membrane», a circular outer domain-containing adhesion structures, is crucial for the tight sealing of the bone area to be resorbed [1]. These adhesions are called podosomes [18] and are constituted by dynamic actin microfilaments, actin-binding proteins and signalling molecules, which move to the periphery of the osteoclast forming a podosomal belt [19]. A further step of cytoskeletal rearrangement before starting resorption is the gathering of podosomes in hooplike structures named actin rings

[19]. Finally, the tight sealing is guaranteed by the integrin receptors, mainly αVβ3 and, to a lesser extent, α2β1 and αVβ5, which ensure the tight anchorage of microfilaments with the extracellular matrix.

The portion of the osteoclast membrane facing the vascular compartment represents the basolateral domain [20], which again participates to the bone resorption function since it is rich of molecules involved in ion transport, and in the response to extracellular stimuli. Just opposite to the ruffled border, the basolateral membrane displays the functional secretory domain [21] that contributes to the release of the bone degradation products into the circulation, through intense vesicular trafficking and transcytosis processes [22].

Bone resorption is a step-by-step process (◘ Fig. 3.2). After adhesion, osteoclasts dissolve first the inorganic components of the bone matrix. To this aim, the carbonic anhydrase II (CAII) accelerates the hydration of carbonic anhydride (CO_2) into carbonic acid (H_2CO_3), which spontaneously dissociates in bicarbonate (HCO_3^-) and proton (H^+) ions. The latter are actively transported in the extracellular microenvironment underneath the cells, called resorption lacuna, by means of a specialised vacuolar-type proton (H^+)-ATPase located in the ruffled border, while the HCO_3^- is exchanged with chloride (Cl^-) through the $HCO3^-/Cl^-$ anion exchanger 2 (AE2) [23, 24]. The Cl^- ions are then moved in the resorption lacuna by a $2Cl^-/1H^+$ antiporter, and the result is the presence of hydrochloric acid (HCl) in the lacuna. This acidic microenvironment dissolves the

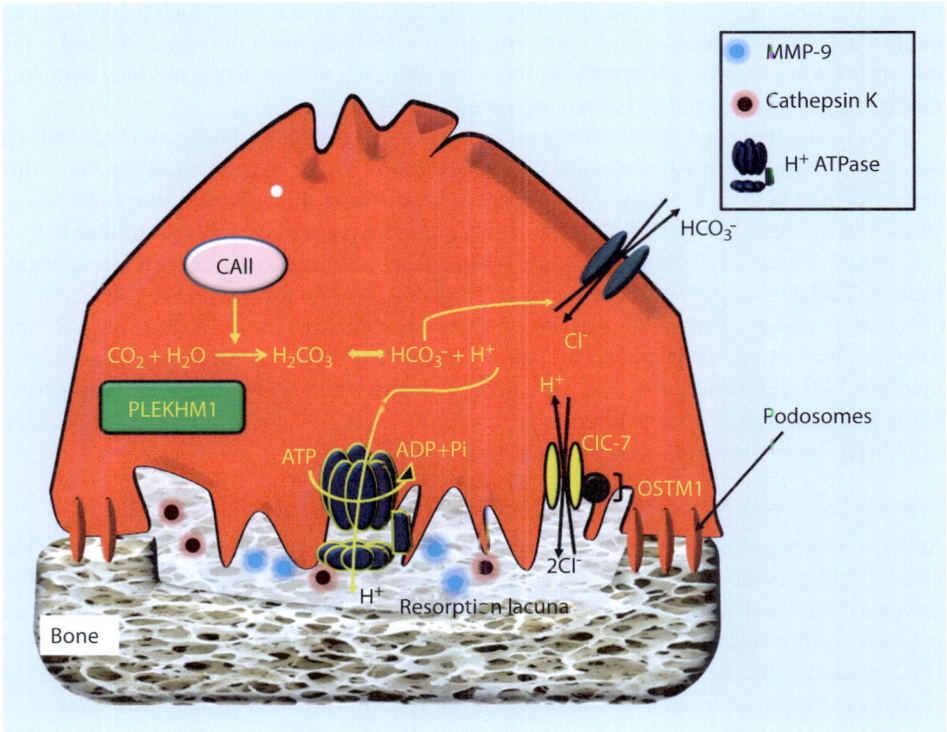

◘ Fig. 3.2 Osteoclast bone resorption. Schematic diagram showing the molecular machinery of bone resorption

hydroxyapatite, exposing the organic bone matrix, which can now be digested by proteolytic enzymes, including cathepsin K [25], released by lysosomal exocytosis. Finally, debris deriving from the digested matrix is removed by the osteoclast through the functional secretory domain by transcytosis [26].

The mechanism of bone resorption has recently been further enriched by other molecules (◘ Fig. 3.2). Pleckstrin homology domain-containing family M (with RUN domain) member 1 (PLEKHM1) [27] is likely involved in vesicular trafficking, while osteopetrosis-associated transmembrane protein 1 (OSTM1) [28] represents the β subunit of the $2Cl^-/1H^+$ antiporter, ensuring its correct placing in the lysosomal and ruffled border membranes.

Osteoclast Regulation of Osteoblasts

It is well known that a close crosstalk between osteoclasts and osteoblasts is crucial to maintain a correct balance between resorption and formation in the bone remodelling process. Although the paracrine regulation of osteoclasts by osteogenic cells is well described, the reciprocal regulation has become apparent only recently.

Definitely, osteoclasts concur to regulate osteoblast formation and recruitment at the sites of bone remodelling through the release of factors stored in the bone matrix, such as transforming growth factor (TGF) β, insulin-like growth factor 1 (IGF1) and bone morphogenetic proteins (BMPs) [29, 30], which recruit and activate osteoblasts in the resorbed area. In addition, what emerged recently is that osteoclasts directly regulate osteoblast differentiation by secreting coupling factors, collectively called clastokines [31–33]. This was previously suggested by the observation that transgenic mice in which osteoclast formation is severely affected present with impaired osteoblast function and decreased bone mineralisation [34, 35]. Conversely, in the osteoclast-rich osteopetrosis models, bone formation rate is not affected or is even increased [36, 37].

One of the first clastokines identified so far is sphingosine 1-phosphate (S1P), which was found to induce in vitro osteoblast differentiation [38, 39]. Consistently, in vivo studies performed with myeloid-specific cathepsin K knockout mice showed that their osteoclasts had increased levels of S1P with a consequent increase of osteoblast number [31].

TRAcP is another evoked clastokine, which likely promotes osteoblast differentiation. Indeed, TRAcP-overexpressing mice have an increased bone formation rate [40]. Similar effects were observed with the collagen triple repeat containing 1 (CTHRC1) [41, 42] and the complement factor 3a (C3a), the latter recently identified in osteoclast-conditioned medium [43]. Interestingly, it has been observed that while CTHRC1 deletion in osteoblasts does not induce a bone phenotype, its conditional knockout in osteoclasts resulted in reduced bone mass and bone formation rate [44].

The Osteoclast Niche: Regulation of Haematopoietic Stem Cells

Another recently discovered function of osteoclasts is the regulation of haematopoiesis, which highlights the high versatility of this cell. An indirect role of HSC regulation for the osteoclast has been ascertained by the means of the MMP-9 and cathepsin K enzymes released during resorption, which regulate the activation of some cytokines crucial for HSC homeostasis. In particular, cathepsin K cleaves the CXC chemokine ligand 12 (CXCL12), responsible for the anchorage of the HSC to the niche, causing the mobilisation of immature haematopoietic progenitor cells [45]. Likewise, MMP-9 allows the release of soluble Kit-ligand (sKit-L), thus promoting the transfer of HSC from the quiescent to proliferative niche. Consistently, in MMP-9 knockout mice, both sKit-L release and HSC motility are impaired, resulting in the failure of haematopoietic recovery [46]. Interestingly, it has been observed that increased levels of local calcium, which could derive from the osteoclast bone-resorbing activity, promote HSC engraftment at the endosteal niche [47].

Other clues suggesting that osteoclasts could regulate HSC rely on the evidence that some treatments affecting osteoclasts also influence HSC homeostasis and vice versa. As an example, bisphosphonates augmented peripheral HSC numbers after their mobilisation with granulocyte colony-stimulating factor (G-CSF) [48], while Sim et al. showed that alendronate increased the long-term engraftment potential and stress resistance of HSCs [49]. Strontium ranelate, which also inhibits osteoclast function, delayed the recovery of HSC after bone marrow transplantation [50]. Likewise, bone marrow HSCs are increased after prostaglandin E2 (PGE2) administration [51]. Finally, mobilisation of HSC after G-CSF treatment is further increased in mice carrying osteopetrosis due to mutation of genes involved in osteoclast differentiation, including M-CSF, c-Fos and RANKL, while it is reduced in osteoprotegerin (OPG) knockout mice [52], characterised by high numbers of osteoclasts.

3.2 Osteoclast Deregulation and Related Pathologies

The picture of osteoclast physiology has undoubtedly benefited from the study of diseases in which their formation and function are deregulated. In fact, the correct balance between osteogenic functions and osteoclast activity is mandatory to preserve the bone and prevent skeletal diseases. Herein we will describe three main pathologies due to the failure of osteoblast-osteoclast coupling.

3.2.1 Osteoporosis

This is a systemic and progressive bone disease characterised by a decrease in bone mass and density, eventually leading to a higher risk of fracture [53, 54]. It has been estimated that approximately 30% of postmenopausal women develop osteoporosis in the United States and in Europe, and at least 40% of these women and 15–30% of men will experience one or more fractures in their remaining lifetime [55, 56]. Furthermore, increase of life expectancy worldwide will be responsible for a steadily increase in the incidence of this disease in the years to come.

The term «primary» osteoporosis refers to a condition related to elderly people and is further classified as type I (i.e. postmenopausal) and type II (age related) [57]. This condition is generally due to an exacerbated osteoclast activity that cannot be compensated by a suitable deposition of new bone by osteoblasts. As a matter of fact, age-related oestrogen and androgen withdrawal is the main guilty of bone mass *defaillance*, since these hormones physiologically act on two fronts: promoting osteoblast survival and function to one side and restraining bone resorption by favouring osteoclast apoptosis to the other [58, 59]. Oestrogens also reduce osteoblast production of pro-osteoclastogenic cytokines, such as tumour necrosis factor (TNF)-α, interleukin (IL)-1β, IL-6 and RANKL, and increase the secretion of OPG [60, 61].

Secondary osteoporosis includes a broad range of osteoporotic features arising from a number of chronic diseases with the onset at any ages [62]. Indeed, bone mass loss could be secondary to the four following disease conditions:

- Endocrine diseases, such as hypogonadism and, to a lesser extent, hyperthyroidism. Glucocorticoid treatment, which is often indicated in the anti-inflammatory therapies, is also responsible for bone loss [63].
- Environmental and lifestyle factors, which include sedentary life, alcohol and use of drugs [64].
- Chronic inflammatory diseases, such as the rheumatoid arthritis [65].
- Reduced mobility as consequence of cerebrovascular accident, spinal cord injury and weightlessness, the latter condition experienced by astronauts living for months at 0 gravity [66, 67].

According to the knowledge of the onset of osteoporosis, animal models have been developed in order to experimentally mimic this disease. With regard to the oestrogen withdrawal-induced osteoporosis, the best model employed is the ovariectomy of adult (age 4–8 weeks old) female mice and rats [68]. Ovariectomized (OVX) animals show a dramatic decrease of trabecular bone mass along with an increase of the osteoclast numbers and of the serum levels of the bone resorption marker carboxy-terminal collagen crosslinks (CTX) [69].

A successful in vivo model of secondary osteoporosis is the hindlimb suspension [70]. This model mimics the bone loss induced by mechanical unloading. Mice or rats are subjected to hindlimb suspension by means of their tail, which is hanged to a swivel apparatus (approximately 30° angle), thus allowing animals to move freely into the cage using their forelimbs and to readily access food and water. After 21 days of suspension, animals present a decrease in their bone volume, due to an increase of osteoclast numbers [71].

Finally, another model useful to mimic disuse osteoporosis is the botulin toxin A (Botox) treatment, which consists in the injection of Botox (2.0 unit/100 g) into the right quadriceps and the posterior compartment of the right calf (targeting gastrocnemius, plantaris and soleus) [72]. Therefore, this treatment induces a transient and local paralysis, eventually leading to hindlimb bone loss, which becomes overt after 21 from treatment [71, 72].

3.2.2 Bone Metastases

Bone metastases represent the fatal destiny of several oncologic patients, especially those affected by breast and prostate carcinomas, in which the incidence of relapse in bone can reach 70 and 90%, respectively [73]. Once bone metastases come up, the chance of survival dramatically drops, and the quality of life deteriorates, eventually leading to a severe morbidity characterised by pain, fractures, nerve compression and, not least, hypercalcemia, due to exacerbated bone resorption [74]. From the clinical and radiographic points of view, bone metastases can be classified into (i) osteosclerotic, in which there is an abnormal deposition of a woven bone, very poor in quality; (ii) osteolytic, due to prominent bone resorption; and (iii) mixed, in which both features coexist in the same metastatic site [75]. Osteolytic lesions are most frequently observed in breast cancer patients, and as suggested by the name, bone erosion is extensive, allowing the tumour cells to create a physical space into a hard tissue, where they can survive and proliferate [75]. Therefore, this pathological condition typically evokes an exacerbated osteoclast activity. As a matter of fact, there is a general consensus about the fact that tumour cells are not able per se to resorb the bone matrix, while they can produce factors that directly and/or indirectly stimulate osteoclast formation and activity, thus incepting the so-called vicious cycle [75]. Several studies tried to picture this complex and tight crosstalk between osteoclasts and cancer cells, paving the way for the identification of new therapeutic strategies able to affect both cell types [76, 77]. Most of these studies relied on the possibility to reproduce the bone metastatic disease in mouse models, through the intracardiac injection of osteotropic tumour cells. This technique, developed by Arguello [78] and implemented by Yoneda [79], implies the injection of tumour cells that have a specific propensity to colonise the bone, into the left ventricle of 4-week-old female mice the latter being immunocompromised if the tumour cells are of human origin. This allows cells to spread in the systemic circulation and colonise the bone. Generally, after 3–4 weeks from cell inoculation, it is possible to appreciate the presence of osteolytic lesions in the hindlimbs of mice by X-ray analysis [80]. Further processing of the hindlimbs for histochemical and histomorphometric analysis allows to determine other features of the bone metastases, which include the increase of osteoclast number and surface over the bone surface by histomorphometric analysis, and the worsening of the tumour burden, by histological staining with haematoxylin and eosin or by immunohistochemistry of specific tumour markers [80].

Among the factors involved in the fuelling of the vicious circle, the parathyroid hormone-related protein (PTHrP) was one of the first spotted protagonists, proven to exert a key role in the development of bone metastases [81]. Moreover, its production by tumour cells is further stimulated by TGFβ [82]. In turn, PTHrP induces osteoblasts and stromal cells to produce RANKL, thus promoting osteoclastogenesis [83].

Other pro-osteolytic factors produced by tumour cells in the bone microenvironment are IL-6, IL-8 and IL-11 [84, 85], cyclooxygenase-2 (COX-2) [86], hypoxia inducible factor (HIF)-1α [87] and TNF-α, all promoting osteoclast formation and resorption. Moreover, a recent study from Sethi et al. [88] demonstrated that tumour-derived Jagged promotes osteolytic bone metastases by the activation of the NOTCH pathway in osteoclasts [88].

What further consolidates the vicious circle is that osteoclasts, while destroying the bone matrix, release and activate several growth factors therein stored, such as TGFβ, BMPs, IGF-1, VEGF and PDGF, which support tumour cell survival and growth. Therefore, it is conceivable that, in order to fight the vicious circle, it is necessary to act on two fronts: (i) inhibit osteoclast activity by antiresorptive therapy and (ii) block local tumour growth, thus counteracting this dangerous synergy.

3.2.3 Osteopetrosis

This pathology features the other side of the coin, that is, when osteoclasts do not work at all. In fact, osteopetrosis (i.e. marble bone disease) is an onomatopoeic term to describe a rare genetic disorder characterised by increased bone density at radiography, now known to be due to the impairment of osteoclast function [89]. The first description of the clinical features of osteopetrosis came from Albers-Schönberg in 1904, who also gave the name to this pathology [90].

Osteopetrosis is a very heterogeneous disease in terms of severity, ranging from asymptomatic to fatal in infancy. Four forms of osteopetrosis are currently classified: autosomal dominant, autosomal recessive, intermediate autosomal recessive and X-linked osteopetrosis [2]. The hallmarks for all these variants are the increase of bone mass, eventually leading to frequent fractures due to a poor quality of the bone, and reduced skull cavities, nerve foramina and bone marrow space. Extra skeletal symptoms are also present, such as anaemia, pancytopenia and hepatosplenomegaly, all due to bone marrow failure [89]. The autosomal recessive osteopetrosis is generally the most severe form, with an incidence of up to 1 in 250,000 births. It is also known as malignant infantile osteopetrosis, due to the lethal outcome and the early onset in the first year of life. It is characterised by dense and fragile bones, deformities, short stature, deafness and blindness, bone marrow failure and impaired immune function, with sepsis and secondary infections and, in some cases, mental retardation due to primary neurodegeneration [91].

Osteopetrosis is a typical osteoclast failure disease, and depending on whether it is due to a defect in osteoclast formation or in osteoclast function, it is classified as osteoclast-rich or osteoclast-poor subtypes. The former is the most common form. In this case, osteoclasts are generated normally, but they do not degrade bone due to loss-of-function mutations in genes encoding for key factors responsible for bone resorption, such as the *TCIRG1*, which encodes the a3 subunit of the H^+-ATPase pump, accounting for the 50% of autosomal recessive cases [92, 93]. Mutations in the *CLCN7* and the *CA II* genes, coding for the $2Cl^-/1H^+$ antiporter and the carbonic anhydrase type II, respectively, are also implicated in this disease [94, 95]. Osteopetrosis due to *CAII* mutations is characterised also by renal tubular acidosis and has an intermediate severity [95]. Loss-of-function mutations of the CLCN7 gene cause a severe autosomal recessive form characterised by lysosomal storage diseases often leading to primary neuropathy [94], while heterozygous missense mutations cause autosomal dominant osteopetrosis, characterised by a milder course.

Other genes recently found to be involved in osteopetrosis are OSTM1, which has the likely role to act as a β-subunit to stabilise the ClC-7 protein [96], and *sorting nexin 10* (*SNX10*), whose product has been suggested to interact with the proton pump [97]. PLEKHM1 is another protein whose deficiency induces an intermediate form of osteopetrosis, due to a defective vesicular trafficking in osteoclasts, eventually leading to impairment of their activity [27].

With regard to the osteoclast-poor type, Sobacchi et al. [98] found in 2007 that mutations in the *TNFSF11* gene, coding for RANKL, caused autosomal recessive osteopetrosis characterised by a complete absence of osteoclasts. Unfortunately, patients carrying this mutation cannot be treated with HSC transplantation, the only therapeutic option for the management of infantile osteopetrosis, because the molecular defect is not intrinsic in the osteoclast but affects RANKL-producing cells, including osteoblasts and stromal cells. Consistently, another study identified mutations in the *TNFRSF11A* gene coding for RANK, giving rise to an autosomal recessive form with a phenotype similar to that caused by RANKL mutations [99] but treatable with HSC transplantation given that this defect is osteoclast autonomous.

3.3 Osteoclast Pharmacology

When osteoclast bone resorption exceeds osteoblast bone formation, a net loss of bone mass occurs. Therefore, these cells are targeted by various drugs in order to rescue a balanced bone turnover. The most prevalent treatments are herein described.

3.3.1 Bisphosphonates

Bisphosphonates include a group of compounds that strongly inhibit osteoclast function, and over the past two decades, they have been largely employed to treat osteoporosis and bone metastases [100]. What made bisphosphonates a good therapeutic strategy is their selective affinity for the bone, due to their P-C-P backbone structure and ability to chelate calcium ions, thus binding to exposed bone mineral and being internalised by bone-resorbing osteoclasts that then undergo apoptosis [101]. Indeed, the simplest bisphosphonates, clodronate and etidronate, employed since 1989, induce massive osteoclast apoptosis because they are incorporated into non-hydrolysable analogues of adenosine triphosphate. The more powerful next-generation nitrogen-containing bisphosphonates (pamidronate, alendronate, ibandronate, risedronate, zoledronate) instead inhibit a key enzyme of the mevalonate pathway, the farnesyl pyrophosphate (FPP) synthase, thus resulting in the accumulation of the isopentenyl diphosphate, which is incorporated into the toxic nucleotide metabolite ApppI, eventually leading to osteoclast apoptosis as well [101]. Moreover, bisphosphonates prevent the prenylation of small GTPases, hereby disrupting the production of isoprenoid lipids in the mevalonate pathway and affecting osteoclast function [102].

With regard to the therapeutic application in the treatment of postmenopausal osteoporosis, while the first generation of bisphosphonates showed a moderate effect on bone resorption, the second- and third-generation compounds were much more potent, as shown by several clinical trials performed during the last 15 years, with a reduction of fracture rates up to 50% [103]. However, recently, an association between long-term bisphosphonate use and atypical femoral fracture risk has emerged [55, 104]. Moreover, due to the high affinity of bisphosphonates for the bone mineral, they accumulate in the bone matrix in long-term treatments, «freezing» the bone and causing it to become much more static [105].

The use of bisphosphonates in the management of bone metastases has beneficial effects not only for osteolytic lesions but also for osteosclerotic metastases, efficiently counteracting skeletal morbidity and improving the quality of life [106]. However, benefits in the increase overall survival of patients need to be more deeply ascertained, and likely as a consequence of the massive dose administered in these patients, some of them develop, as an adverse effect, the osteonecrosis of the jaw [107].

3.3.2 Denosumab

Given the pivotal role of the RANKL/RANK/OPG axis in the biology of the osteoclast, many therapeutic strategies were diverted to block this pathway. A recent drug developed for this purpose is denosumab, a human monoclonal IgG2 antibody specifically raised against the soluble and cell surface RANKL, thus inhibiting its binding to RANK with a resulting block of osteoclastogenesis [108]. Overall, the effectiveness of this drug in postmenopausal osteopenic women seems to be not inferior, or even greater, than that of the reference drug alendronate, with a better patient compliance, since administration is performed subcutaneously once every 6 months [109–111]. Denosumab treatment also improves bone mechanical properties [112], and its effect on bone remodelling seems to be reversible [113] after stopping the treatment, thus avoiding a permanent loss of dynamicity of the bone. Finally, it has been reported that denosumab has the same adverse effect of alendronate [114], while other studies found hypocalcaemia in few treated patients [109].

3.3.3 New Antiresorptive Agents

Due to the possible side effects observed in the currently available antiresorptive therapies, the need to identify new targets for alternative treatments is still present [115]. In this respect, the most recently identified strategies are described below.

Cathepsin K Inhibitors

Cathepsin K belongs to a large family of at least 11 cysteine proteases in humans. As already mentioned, it is highly expressed and released by osteoclasts inside the resorption lacuna where it breaks down type I collagen. Indeed, the pivotal role of this enzyme in osteoclasts has been clearly demonstrated by the fact that mutations of its gene lead to pycnodysostosis (i.e. Toulouse-Lautrec syndrome), a rare genetic disease characterised by an impairment of bone organic matrix resorption, while osteoclasts form and demineralise the bone matrix normally [116, 117]. From a therapeutic point of view, the fact that cathepsin K inhibitors block osteoclast activity and do not prevent osteoclast formation is a remarkable advantage, given that it allows the osteoclast to perform other physiologic functions, one for all the regulations of osteoblasts. Odanacatib (MK-0822) is a cathepsin K inhibitor that has been employed in phase I and II clinical trials in postmenopausal women, proving to be effective in preventing bone resorption [118]. This compound has also been found to reduce serum bone resorption biomarkers in breast cancer patients with bone metastases [119]. Another cathepsin K inhibitor under phase I and II clinical trial is ONO-5334, which shows an antiresorptive activity similar to that of bisphosphonates [120].

αVβ3 Integrin Inhibitors

αVβ3 integrin is also a potential antiresorptive and antitumoural target, since this receptor participates in osteoclast adhesion, triggering a complex intracellular signalling involving tyrosine kinases c-Src and Syk [121], a crucial step for the assessment of bone resorption. αVβ3 integrin inhibitors (i.e. S247, ATN-161, PSK1404) have been mainly employed in preclinical trials to treat bone metastases, providing the double advantage to target osteoclasts and tumour cells, since in the latter this integrin can be highly expressed and provide a pro-invasive phenotype [122]. L-000845704 is another αVβ3 antagonist already employed in clinical trials, which proved to inhibit bone resorption in women with postmenopausal osteoporosis [123].

3.4 Methods for Osteoclast Primary Cultures

First attempts to obtain osteoclast primary cultures, dating back more than 30 years ago, include mechanical disaggregation of the long bones of newborn animals (i.e. rabbits and rats), which provided short living mature osteoclasts [124, 125]. Successful mature osteoclasts were also isolated from the long bones of laying hens kept in low-calcium diet, a condition that enhances osteoclast formation and activity [126].

The identification of M-CSF and RANKL as powerful pro-osteoclastogenic factors has greatly contributed to the generation of osteoclast primary cultures from different sources. In fact, nowadays, the preferential way to obtain primary osteoclasts is by isolating osteoclast precursors from the bone marrow or peripheral blood mononuclear cells and differentiating them using these two pro-osteoclastogenic cytokines. This method results in a large number of purified differentiated osteoclasts.

3.4.1 Osteoclast Isolation from Mouse Bone Marrow Cells

The fulfilment of this strategy has been supported by the ascertainment that osteoblasts and stromal cells inside the bone marrow produce a plethora of factors that positively regulate osteoclast differentiation from the monocyte/macrophage lineage. As described below, two main protocols can be used to obtain osteoclasts from bone marrow cells, one mimicking an osteoblast-preosteoclast coculture and the other requiring the purification of osteoclast precursors from the bone marrow.

Osteoclast Primary Culture from Unfractionated Bone Marrow Cells

This method starts from a bone marrow culture, which contains osteoclast precursors and mesenchymal stromal cells supporting osteoclastogenesis by producing RANKL and M-CSF under the stimulation of 1α,25-dihydroxy-vitamin D_3 [1,25(OH)$_2D_3$] added to the culture [127, 128]. Seven-day-old mice are euthanized by CO_2 inhalation, and then the fore- and hindlimbs are excised and put in HANK's balanced salt solution (HBSS). A gross cleaning with a blade is performed to remove the surrounding soft tissues; hence, the long bones are finely minced into small pieces, and bone marrow is mechanically flushed out by repeated pipetting with a Pasteur pipette. Collected bone marrow cells are cultured in Dulbecco's modified eagle medium (DMEM) plus 10% foetal bovine serum (FBS); then, the day after, medium is replaced by fresh DMEM supplemented with 10% FBS and 10^{-8} M 1,25(OH)$_2D_3$. After 8–10 days of culture, multinucleated osteoclasts appear and can be subjected to characterisation [69].

Osteoclast Primary Culture from Purified Bone Marrow Macrophages

This method requires the isolation of osteoclast precursors from the bone marrow and their subsequent differentiation by treatment with M-CSF and RANKL. Briefly, the bone marrow is flushed out from the limbs of 7-day-old mice as described above; then it is diluted 1:1 in HBSS and stratified over Ficoll/Histopaque 1077. After centrifugation at 400 g for 30 min without brake, the solution appears stratified in three layers. Osteoclast precursors are recovered from the white ringlike intermediate layer, resuspended in DMEM containing 10% FBS and plated. After 3 h, cultures are washed to remove non-adherent cells, and then DMEM supplemented with 10% FBS, 50 ng/ml mouse recombinant (mr) M-CSF and 50 ng/ml mrRANKL are added. After 5–7 days of culture, it is possible to observe an enriched multinucleated osteoclast population [69].

3.4.2 Osteoclast Isolation from Human Peripheral Blood

The setup of this primary culture has greatly contributed to the study of the biology of osteoclasts in normal and pathological conditions, allowing to recover osteoclasts from an available human source, the peripheral blood, instead of bone marrow biopsies [93, 129].

Human peripheral blood mononuclear cells are obtained by density gradient of fresh peripheral blood [93]. This sample is first diluted 1:1 with PBS or HBSS and then stratified over Ficoll/Histopaque 1077 and centrifuged at 800 g for 30 min with the brake off. This procedure will again result in three layers, of which the middle white contains the osteoclast precursors. Isolated buffy coat cells are then washed with HBSS and centrifuged and resuspended in DMEM plus 10% FBS and plated. After 3 h, plates are rinsed to remove non-adherent cells and cultured in the aforementioned medium in the presence of 50 ng/ml human recombinant (hr) M-CSF and 30 ng/ml hrRANKL. The culture requires at least 2 weeks in order to provide multinucleated osteoclasts and 3 weeks to assess bone resorption, during which the medium should be replaced every 3–4 days.

3.4.3 Evaluation of Osteoclast Differentiation

Three main determinants demonstrate that bona fide osteoclasts have been formed in the primary culture: (1) the presence of giant multinucleated cells (more than three nuclei/cell) which can be easily observed by phase contrast microscopy, (2) the positivity of these cells to the osteoclast marker TRAcP and (3) their ability to resorb the bone. This latter will be discussed in the next paragraph.

Histochemical TRAcP Assay

As suggested by the name, TRAcP belongs to a class of metalloenzymes that catalyse the hydrolysis of various phosphate esters and anhydrides under acidic conditions. Although it has always been considered a classical marker of osteoclasts, TRAcP is also expressed by inflammatory macrophages and dendritic cells. TRAcP activity can be easily evaluated in osteoclast cultures fixed in 4% paraformaldehyde, by a histochemical assay using a commercially available kit (Sigma-Aldrich #387A) according to the manufacturer's instruction. TRAcP-positive osteoclasts appear as purple stained cells with three or more nuclei (◘ Fig. 3.3a, b).

Transcriptional Evaluation of Osteoclast-Specific Genes

As described above, during osteoclast differentiation, the activated NF-kB transcription factor promotes the transcriptional expression of downstream osteoclast-specific genes, which can be evaluated by RT-PCR. Typical genes whose expression increases during

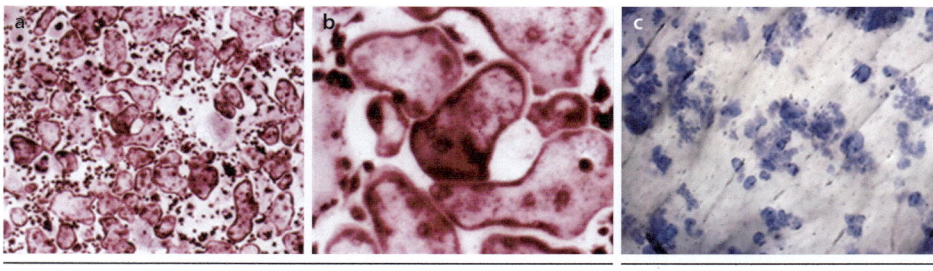

TRAcP   Toluidine blue

◘ Fig. 3.3 a Picture showing an osteoclast primary culture obtained from purified bone marrow macrophages and subjected to cytochemical assay for the tartrate-resistant acid phosphatase (TRAcP) activity (magnification, 25×). b Magnification of a (100×). c Picture of a bone slice stained with toluidine *blue* showing several resorption pits evidenced as *blue* spot (magnification = 100×)

osteoclastogenesis are cathepsin K, CTR, MMP-9, RANK and TRAcP [69]. Genes whose expression is associated with preosteoclast fusion can also be evaluated, such as DC-STAMP, CD44 and macrophage fusion receptor (MFR).

3.4.4 Evaluation of Osteoclast Function

In vitro osteoclast activity can be easily assessed by the resorption pit assay. Osteoclast precursors are plated on dentine/bone slices (commercially available) and differentiated into mature osteoclasts as described above. Alternatively, after their differentiation in plastic dishes, mature osteoclasts are detached and replated onto dentine/bone slices for at least 48 h [66]. Osteologic dishes covered with a layer of hydroxyapatite can also be used. In both cases, at the end of the resorption period (48 h for rodent osteoclasts and 1 week for human osteoclasts), cells are mechanically removed from the bone surface, and sections are stained with toluidine blue or Coomassie blue or observed by scanning electron microscopy. Toluidine blue and Coomassie blue staining are very easy procedures, which exploit the higher affinity of these dyes for the organic bone matrix components that are exposed after removal of the mineral by the action of osteoclasts. These areas appear more intensely stained, and in the case of toluidine blue, they are metachromatic for the content of glycosaminoglycans (◘ Fig. 3.3c). Scanning electron microscopy is more time-consuming but can provide more detailed information on pit shape, size and depth. Total pit area and volume are then quantified by light and by scanning electron microscopy, respectively, using a morphometric software. Alternatively, pits are classified in small (<10 μm diameter), medium (10–30 μm diameter) and large (>30 μm diameter), and their number is multiplied for a size score of 0.3 for small pits, 1 for medium pits and 3 for large pits. The sum of the three scores provides the final pit index that is proportional of the pit number and size [130]. Other methods make use of dentin/bone sections pre-stained with calcium-binding fluorofores (calcein, alizarin red or inactive fluorescent bisphosphonates). Fluorofores are removed during bone resorption, and in this case, the resorption pits appear as dark areas in a fluorescent background. They are then evaluated by fluorescence or confocal microscopy using the same parameters described above.

3.5 Conclusions

Osteoclasts are bone-resorbing cells that largely contribute to bone remodelling. Their deregulated activity impacts the bone health, and for this reason, they are considered important targets for therapy. Osteoclast biotechnology has largely furthered the knowledge on their biology and added to the development of therapies to block their exacerbated activity (i.e. in osteoporosis, bone metastases) or, conversely, re-establish their impaired formation or resorption (i.e. osteopetrosis). More is expected in the years to come to broaden the impact of their therapeutic management in other pathological conditions that currently have no cure.

> **Take-Home Messages**
>
> - The osteoclast is a monocyte/macrophage arising cell whose differentiation is a fine-tuned process regulated by systemic and local factors, among which the RANKL/RANK pathway plays a crucial role.
> - The osteoclast is more than a hungry cell that eats bone, showing a versatile profile in terms of functions, which include the regulation of osteoblastogenesis and haematopoiesis.
> - Several skeletal diseases are the result of a dysfunction of osteoclast formation and activity, which makes this cell a useful target for therapy.
> - The investigation of osteoclast pathophysiology can rely on well-established in vitro methods of precursor isolation from mouse bone marrow and human peripheral blood.

References

1. Cappariello A, Maurizi A, Veeriah V, Teti A. *The Great Beauty* of the osteoclast. Archiv Biochem Biophys. 2014;558:70–8.
2. Del Fattore A, Cappariello A, Teti A. Genetics, pathogenesis and complications of osteopetrosis. Bone. 2008;42(1):19–29.
3. Walker DG. Bone resorption restored in osteopetrotic mice by transplant of normal bone marrow and spleen cells. Science. 1975;190(4216):784–5.
4. Rasmussen H, Bordier P. The cellular basis of metabolic bone disease. New Engl J Med. 1973;289(1):25–32.
5. Chambers TJ. The birth of the osteoclast. Ann N Y Acad Sci. 2010;1192:19–26.
6. Tondravi MM, McKercher SR, Anderson K, Erdmann JM, Quiroz M, Maki R, et al. Osteopetrosis in mice lacking hematopoietic transcription factor PU.1. Nature. 1997;386(6620):81–4.
7. Van de Wijngaert FP, Tas MC, Burger EH. Conditioned medium of fetal mouse long bone rudiments stimulates the formation of osteoclast precursor-like cells from mouse bone marrow. Bone. 1989;10(1):61–8.
8. Menaa C, Kurihara N, Roodman GD. CFU-GM-derived cells from osteoclasts at a very high efficiency. Biochem Biophys Res Commun. 2000;267(3):943–6.
9. Biskobing DM, Fan X, Rubin J. Characterization of MCSF-induced proliferation and subsequent osteoclast formation in murine marrow culture. J Bone Miner Res. 1995;10(7):1025–32.
10. Simonet WS, Lacey DL, Dunstan CR, Kelley M, Chang MS, Luthy R, et al. Osteoprotegerin: a novel secreted protein involved in the regulation of bone density. Cell. 1997;89(2):309–19.
11. Lacey DL, Timms E, Tan HL, Kelley MJ, Dunstan CR, Burgess T, et al. Osteoprotegerin ligand is a cytokine that regulates osteoclast differentiation and activation. Cell. 1998;93(2):165–76.
12. Yasuda H, Shima N, Nakagawa N, Yamaguchi K, Kinosaki M, Mochizuki S, et al. Osteoclast differentiation factor is a ligand for osteoprotegerin/osteoclastogenesis-inhibitory factor and is identical to TRANCE/RANKL. Proc Natl Acad Sci U S A. 1998;95(7):3597–602.
13. Dougall WC, Glaccum M, Charrier K, Rohrbach K, Brasel K, De Smedt T, et al. RANK is essential for osteoclast and lymph node development. Genes Dev. 1999;13(18):2412–24.
14. Teitelbaum SL. Osteoclasts: what do they do and how they do it? Am J Pathol. 2007;170(2):427–35.
15. Franzoso G, Carlson L, Xing L, Poljak L, Shores EW, Brown KD, et al. Requirement for NF-kappaB in osteoclast and B-cell development. Genes Dev. 1997;11(24):3482–96.
16. Kukita T, Wada N, Kukita A, Kakimoto T, Sandra F, Toh K, et al. RANKL-induced DC-STAMP is essential for osteoclastogenesis. J Exp Med. 2004;200(7):941–6.
17. Nakashima T, Hayashi M, Takayanagi H. New insights into osteoclastogenic signaling mechanisms. Trends Endocrinol Metab. 2012;23(11):582–90.
18. Marchisio PC, Cirillo D, Naldini L, Primavera MV, Teti A, Zambonin Zallone A. Cell-substratum interaction of cultured avian osteoclasts is mediated by specific adhesion structures. J Cell Biol. 1984;99(5):1696–705.

19. Saltel F, Chabadel A, Bonnelye E, Jurdic P. Actin cytoskeletal organisation in osteoclasts: a model to decipher transmigration and matrix degradation. Eur J Cell Biol. 2008;87(8–9):459–68.
20. Supanchart C, Kornak U. Ion channels and transporters in osteoclasts. Arch Biochem Biophys. 2008;47(2):161–5.
21. Vääräniemi J, Halleen JM, Kaarlonen K, Ylipahkala H, Alatalo SL, Andersson G, et al. Intracellular machinery for matrix degradation in bone-resorbing osteoclasts. J Bone Miner Res. 2004;19(9):1432–40.
22. Hirvonen MJ, Fagerlund K, Lakkakorpi P, Väänänen HK, Mulari MT. Novel perspectives on transcytotic route in osteoclasts. Bonekey Rep. 2013;2:306.
23. Teti A, Blair HC, Teitelbaum SL, Kahn A, Koziol C, Konsek J, et al. Cytoplasmic pH regulation and chloride/bicarbonate exchange in avian osteoclasts. J Clin Invest. 1989;83(1):227–33.
24. Lindsey AE, Schneider K, Simmons DM, Baron R, Lee BS, Kopito RR. Functional expression and subcellular localization of an anion exchanger cloned from choroid plexus. Proc Natl Acad Sci U S A. 1990;87(14):5278–82.
25. Henriksen K, Sorensen MG, Nielsen RH, Gram J, Schaller S, Dziegel MH, et al. Degradation of the organic phase of bone by osteoclasts: a secondary role for lysosomial acidification. J Bone Miner Res. 2006;21(1):58–88.
26. Nesbitt SA, Horton MA. Trafficking of matrix collagens through bone-resorbing osteoclasts. Science. 1997;276(5310):266–9.
27. Van Wesenbeck L, Odgren PR, Coxon FP, Frattini A, Moens P, Perdu B, et al. Involvement of PLEKHM1 in osteoclastic vesicular transport and osteopetrosis in incisors absent rats and humans. J Clin Invest. 2007;117(4):919–30.
28. Lange PF, Wartosch L, Jentsch TJ, Fuhrmann JC. ClC7 requires Ostm1 as a beta-subunit to support bone resorption and lysosomal function. Nature. 2006;440(7081):220–3.
29. Xian L, Wu X, Pang L, Lou M, Rosen CJ, Qiu T, et al. Matrix IGF-1 maintains bone mass by activation of mTOR in mesenchymal stem cells. Nat Med. 2012;18(7):1095–101.
30. Pfeilschifter J, D'Souza SM, Mundy GR. Effect of transforming growth factor-beta on osteoblastic osteosarcoma cells. Endocrinology. 1987;121(1):212–8.
31. Lotinun S, Kiviranta R, Matsubara T, Alzare JA, Neff L, Luth A, et al. Osteoclast-specific cathepsin K deletion stimulates S1P-dependent bone formation. J Clin Invest. 2013;123(2):666–81.
32. Teti A. Mechanisms of osteoclast-dependent bone formation. Bonekey Rep. 2013;2:449.
33. Drissi H, Sanjay A. The multifaceted osteoclast; far and beyond bone resorption. J Cell Biochem. 2016;117(8):1753–6.
34. Dai XM, Zong XH, Akhter MP, Stanley ER. Osteoclast deficiency results in disorganized matrix, reduced mineralization, and abnormal osteoblast behavior in developing bone. J Bone Miner Res. 2004;19(9):1441–51.
35. Sakagami N, Amizuka N, Li M, Takeuch K, Hoshino M, Nakamura M, et al. Reduced osteoblastic population and defective mineralization in osteopetrotic (op/op) mice. Micron. 2005;36(7–8):688–95.
36. Del Fattore A, Peruzzi B, Rucci N, Recchia I, Capparello A, Longo M, et al. Clinical, genetic, and cellular analysis of 49 osteopetrotic patients: implication for diagnosis and treatment. J Med Genet. 2006;43(4):315–25.
37. Pennypacker B, Shea M, Liu Q, Masarachia P, Saftig P, Rodan S, et al. Bone density, strength, and formation in adult cathepsin K−/− mice. Bone. 2009;44(2):199–207.
38. Ruy J, Kim HJ, Chang EJ, Huang H, Barno Y, Kim HH, et al. Sphingosine 1-phosphatase as a regulator of osteoclast differentiation and osteoclast-osteoblast coupling. EMBO J. 2006;25 24):5840–51.
39. Pederson L, Ruan M, Westendorf JJ, Khosla S, Ourser MJ. Regulation of bone formation by osteoclasts involves Wnt/BMP signaling and the chemokine sphingosine-1-phosphatase. Proc Natl Acad Sci U S A. 2008;105(52):29764–9.
40. Hayman AR, Cox TM. Tartrate-resistant acid phosphatase knockout mice. J Bone Miner Res. 2003;18(10):1905–7.
41. Kimura H, Kwan KM, Zhang Z, Deng JM, Darnay BG, Behringer RR, et al. Cthrc1 is a positive regulator of osteoblastic bone formation. PLoS One. 2008;3 9):e3174.
42. Takeshita S, Fumoto T, Matsuoka K, Park KA, Aburatani H, Kato S, et al. Osteoclast-secreted CTHRC1 in the coupling of bone resorption and bone formation. J Clin Invest. 2013;123(9):3914–24.
43. Matsuoka K, Park KA, Ito M, Ikeda K, Takeshita S. Osteoclast-derived complement component 3a stimulates osteoblast differentiation. J Bone Miner Res. 2014;29(7):1522–30.
44. Charles JF, Aliprantis AO. Osteoclasts, more than «bone eaters». Trends in Mol Med. 2014;20(8):449–59.

45. Kollet O, Dar A, Shivtiel S, Kalinkovich A, Lapid K, Sztainberg Y, et al. Osteoclasts degrade endosteal components and promote mobilization of hematopoietic progenitor cells. Nat Med. 2006;12(6):657–64.
46. Heissig B, Hattori K, Dias S, Friedrich M, Ferris B, Hackett NR, et al. Recruitment of stem and progenitor cells from the bone marrow niche requires MMP-9 mediated release of kit-ligand. Cell. 2002;109(5):625–37.
47. Adams GB, Chabner KT, Alley IR, Olson DP, Szczepiorkowski ZM, Poznansky MC, et al. Stem cell engraftment at the endosteal niche is specified by the calcium-sensing receptor. Nature. 2006;439(7076):599–603.
48. Takamatsu Y, Simmons PJ, Moore RJ, Morris HA, To IB, Lè Vesque JP. Osteoclast-mediated bone resorption is stimulated during short-term administration of granulocyte colony-stimulating factor but is not responsible for hematopoietic progenitor cell mobilization. Blood. 1998;92(9):3465–73.
49. Sim HJ, Kook SH, Yun CY, Bhattarai G, Cho ES, Lee JC. Brief report: consecutive alendronate administration-mediated inhibition of osteoclasts improves long-term engraftment potential and stress resistance of HSCs. Stem Cells. 2016;34(10):2601–7.
50. Lymperi S, Horwood N, Marley S, Gordon MY, Cope AP, Dazzi F. Strontium can increase some osteoblasts without increasing hematopoietic stem cells. Blood. 2008;111(3):1173–8.
51. Frisch BJ, Porter RL, Gigliotti BJ, Olm-Shipman AJ, Weber JM, O'Keefe RJ, et al. In vivo prostaglandin E2 treatment alters the bone marrow microenvironment and preferentially expands short-term hematopoietic stem cells. Blood. 2009;114(19):4054–63.
52. Miyamoto T. Role of osteoclasts in regulating hematopoietic stem and progenitor cells. World J Orthop. 2013;4(4):198–206.
53. O'Connor KM. Evolution and treatment of osteoporosis. Med Clin North Am. 2016;100(4):807–26.
54. Kruger MC, Wolber FM. Osteoporosis: modern paradigms for last century's bones. Forum Nutr. 2016;8(6):E376.
55. Compston J. Pathophysiology of atypical femoral fractures and osteonecrosis of the jaw. Osteoporos Int. 2011;22(12):2951–61.
56. Sattui SE, Saag KG. Fracture mortality: associations with epidemiology and osteoporosis treatment. Nat Rev Endocrinol. 2014;10(10):592–602.
57. Hendrickx G, Boudin E, Van Hul W. A look behind the scenes: the risk and pathogenesis of primary osteoporosis. Nat Rev Rheumatol. 2015;11(8):462–74.
58. Hughes DE, Dai A, Tiffee JC, Li HH, Mundy GR, Boyce BF. Estrogen promotes apoptosis of murine osteoclasts mediated by TGF-beta. Nat Med. 1996;2(10):1132–6.
59. Xing L, Boyce BF. Regulation of apoptosis in osteoclasts and osteoblastic cells. Biochem Biophys Res Commun. 2005;328(3):709–20.
60. Hofbauer LC, Khosla S, Dunstain CR, Lacey DL, Spelsberg TC, Riggs BL. Estrogen stimulates gene expression and protein production of osteoprotegerin in human osteoblastic cells. Endocrinology. 1999;140(9):4367–70.
61. Shevde NK, Bendixen AC, Dienger KM, Pike JW. Estrogens suppress rankl-induced osteoclast differentiation via a stromal cell independent mechanism involving c-jun repression. Proc Natl Acad Sci U S A. 2000;97(14):7829–34.
62. Diab DL, Watts NB. Secondary osteoporosis: differential diagnosis and workup. Clin Obstet Gynecol. 2013;56(4):686–93.
63. Seibel MJ, Cooper MS, Zhou H. Glucocorticoid-induced osteoporosis: mechanisms, management and future perspectives. Lancet Diabetes Endocrinol. 2013;1(1):59–70.
64. Sapre S, Thakur R. Lifestyle and dietary factors determine age at natural menopause. J Midlife Health. 2014;5(1):3–5.
65. Węgierska M, Dura M, Blumfield E, Żuchowski P, Waszczak M, Jeka S. Osteoporosis diagnostics in patients with rheumatoid arthritis. Rheumatologia. 2016;54(1):29–34.
66. Edwards WB, Schnitzer TJ, Troy KL. The mechanical consequence of actual bone loss and simulated bone recovery in acute spinal cord injury. Bone. 2014;60:141–7.
67. Rucci N, Rufo A, Alamanou M, Teti A. Modeled microgravity stimulates osteoclastogenesis and bone resorption by increasing osteoblast RANKL/OPG ratio. J Cell Biochem. 2007;100(2):464–73.
68. Lelovas PP, Xanthos TT, Thoma SE, Lyritis GP, Dontas IA. The laboratory rat as an animal model for osteoporosis research. Comp Med. 2008;58(5):424–30.
69. Rucci N, Rufo A, Alamanou M, Capulli M, Del Fattore A, Åhrman E, et al. The glycosaminoglycan-binding domain of PRELP acts as a cell type-specific NF-κB inhibitor that impairs osteoclastogenesis. J Cell Biol. 2009;187(5):669–83.

70. Sakata T, Sakai A, Tsurukami H, Okimoto N, Okazaki Y, Ikeda S, et al. Trabecular bone turnover and bone marrow cell development in tail-suspended mice. J Bone Miner Res. 1999;14(9):1596–604.
71. Rucci N, Capulli M, Piperni SG, Cappariello A, Lau F, Frings-Meuthen P, et al. Lipoca in 2: a new mechanoresponding gene regulating bone homeostasis. J Bone Miner Res. 2015;30(2):357–68.
72. Warner SE, Sanford DA, Becker BA, Bain SD, Srinivasan S, Gross TS. Botox induced muscle paralysis rapidly degrades bone. Bone. 2006;38(2):257–64.
73. Kan C, Vargas G, Pape FL, Clézardin P. Cancer cell colonisation in the bone microenvironment. Int J Mol Sci. 2016;17(10):E1674.
74. Coleman RE. Clinical features of metastatic bone disease and risk of skeletal morbidity. Clin Cancer Res. 2006;12(20 Pt 2):6243s–9s.
75. Roodman GD. Mechanisms of bone metastasis. New Engl J Med. 2004;350(16):1655–64.
76. Clézardin P, Teti A. Bone metastasis: pathogenesis and therapeutic implications. Clin Exp Metastasis. 2007;24(8):599–608.
77. Jehn CF, Diel IJ, Overkamp F, Kurth A, Schaefer R, Miller K, et al. Management of metastatic bone disease algorithms for diagnostics and treatment. Anticancer Res. 2016;36(6):2631–7.
78. Arguello F, Baggs RB, Frantz CN. A murine model of experimental metastasis to bone and bone marrow. Cancer Res. 1988;48(3):6876–81.
79. Yoneda T, Sasaki A, Dunstan C, Williams PJ, Bauss F, De Clerck YA, et al. Inhibition of osteolytic bone metastasis of breast cancer by combined treatment with the bisphosphonate ibandronate and tissue inhibitor of the matrix metalloproteinase-2. J Clin Invest. 1997;99(10):2509–17.
80. Rucci N, Recchia I, Angelucci A, Alamanou M, De Fattore A, Fortunati D, et al. Inhibition of protein kinase c-Src reduces the incidence of breast cancer metastases and increases survival in mice: implications for therapy. J Pharmacol Exp Ther. 2006;318(1):161–72.
81. Guise TA, Yin JJ, Taylor SD, Kumagai Y, Dallas M, Boyce BF, et al. Evidence for a causal role of parathyroid hormone-related protein in the pathogenesis of human breast cancer-mediated osteolysis. J Clin Invest. 1996;98(7):1544–9.
82. Yin JJ, Selander K, Chirgwin JM, Dallas M, Grubbs BG, Wieser R, et al. TGF-beta signaling blockade inhibits PTHrP secretion by breast cancer cells and bone metastases development. J Clin Invest. 1999;103(2):197–206.
83. Weilbaecher KN, Guise TA, McCauley LK. Cancer and bone: a fatal attraction. Nat Rev Cancer. 2011;11(7):411–25.
84. Zhang Y, Fujita N, Oh-hara T, Morinaga Y, Nakagawa T, Yamada M, et al. Production of interleukin-11 in bone-derived endothelial cells and its role in the formation of osteolytic bone metastases. Oncogene. 1998;16(3):693–703.
85. Bendre MS, Montague DC, Peer T, Akel NS, Gaddy D, Suva LJ. Interleukin-8 stimulation of osteoclastogenesis and bone resorption is a mechanism for the increased osteolysis of metastastic bone disease. Bone. 2003;33(1):28–37.
86. Singh B, Berry JA, Shoher A, Ayers GD, Wei C, Lucci A. COX-2 involvement in breast cancer metastasis to bone. Oncogene. 2007;26(26):3789–96.
87. Hiraga T, Kizaka-Kondoh S, Hirota K, Hiraoka M, Yoneda T. Hypoxia and hypoxia- inducible factor-1 expression enhance osteolytic bone metastases of breast cancer. Cancer Res. 2007;67(9):4157–63.
88. Sethi N, Dai X, Winter CG, Kang Y. Tumor-derived JAGGED1 promotes osteolytic bone metastases of breast cancer by engaging notch signaling in bone cells. Cancer Cell. 2011;19(2):192–205.
89. Tolar J, Teitelbaum SL, Orchad PJ. Osteopetrosis. N Engl J Med. 2004;351(27):2839–49.
90. Albers-Schönberg HE. Röntgenbilder einer selteren Knock-enerkrankung. Munch Med Wochenscher. 1904;5:365–8.
91. Villa A, Guerrini MM, Cassani B, Pangrazio A, Sobacchi C. Infantile malignant, autosomal recessive osteopetrosis: the rich and the poor. Calcif Tissue Int. 2009;84(1):1–12.
92. Sobacchi C, Frattini A, Orchard P, Porras O, Tezcan I, Andolina M, et al. The mutational spectrum of human malignant autosomal recessive osteopetrosis. Human Mol Genet. 2001;10(17):1767–73.
93. Taranta A, Migliaccio S, Recchia I, Caniglia M, Luciani M, De Rossi G, et al. Genotype-phenotype relationship in human ATP6i-dependent autosomal recessive osteopetrosis. Am J Pathol. 2003;162(1):57–68.
94. Kornak U, Kasper D, Bosl MR, Kaiser E, Schweizer M, Schulz A, et al. Loss of the ClC-7 chloride channel leads to osteopetrosis in mice and man. Cell. 2001;104(2):205–15.
95. Sly WS, Hewett-Emmett D, Whyte MP, Yu YS, Tashian RE. Carbonic anhydrase II deficiency identified as the primary defect in the autosomal recessive syndrome of osteopetrosis with renal tubular acidosis and cerebral calcification. Proc Natl Acad Sci U S A. 1983;80(9):2752–6.

96. Pangrazio A, Poliani PL, Megarbane A, Lefranc G, Lanino E, Di Rocco M, et al. Mutations in OSTM1 (grey lethal) define a particularly severe form of autosomal recessive osteopetrosis with neural involvement. J Bone Miner Res. 2006;21(7):1098–105.
97. Pangrazio A, Fasth A, Sbardellati A, Orchard PJ, Kasow KA, Raza J, et al. SNX10 mutations define a subgroup of human autosomal recessive osteopetrosis with variable clinical severity. J Bone Miner Res. 2013;28(5):1041–9.
98. Sobacchi C, Frattini A, Guerrini MM, Abinun M, Pangrazio A, Susani L, et al. Osteoclast-poor human osteopetrosis due to mutations in the gene encoding RANKL. Nat Genet. 2007;39(8):960–2.
99. Guerrini MM, Sobacchi C, Cassani B, Abinun M, Kilic SS, Pangrazio A, et al. Human osteoclast-poor osteopetrosis with hypogamma-globulinemia due to TNFRSF11A (RANK) mutations. Am J Human Genet. 2008;83(1):64–76.
100. Zhou J, Ma X, Wang T, Zhai S. Comparative efficacy of bisphosphonates in short-term fracture prevention for primary osteoporosis: a systemic review with network meta-analyses. Osteoporosis Int. 2016;27(11):3289–300.
101. Rogers MJ. New insights into the molecular mechanisms of action of bisphosphonates. Curr Pharm Des. 2003;9(32):2643–58.
102. Rogers MJ, Crockett JC, Coxon FP, Mönkkönen J. Biochemical and molecular mechanisms of action of bisphosphonates. Bone. 2011;49(16):34–41.
103. Eriksen EF, Diez-Perez A, Boonen S. Update of long-term treatment with bisphosphonates for postmenopausal osteoporosis: a systematic review. Bone. 2014;58:126–35.
104. Odvina CV, Zerwekh JE, Rao DS, Maalouf N, Gottschalk FA, Pak CY, et al. Severely suppressed bone turnover: a potential complication of alendronate therapy. J Clin Endocrinol Metab. 2005;90(3):1294–301.
105. Pazianas M, van der Geest S, Miller P. Bisphosphonates and bone quality. Bonekey Rep. 2014;3:529.
106. Young RJ, Coleman RE. Zoledronic acid to prevent and treat cancer metastasis: new perspects for an old drug. Future Oncol. 2013;9(5):633–43.
107. Filleul O, Crompot E, Saussez S. Bisphosphonate-induced osteonecrosis of the jaw: a review of 2400 patient cases. J Cancer Res Clin Oncol. 2010;136(8):1117–24.
108. Kostenuik PJ, Nguyen HQ, McCabe J, Warmington KS, Kurahara C, Sun N, et al. Denosumab, a fully human monoclonal antibody to RANKL, inhibits bone resorption and increases BMD in knock-out mice that express chimeric (murine/human) RANKL. J Bone Miner Res. 2009;24(2):182–95.
109. McClung MR, Lewiecki EM, Cohen SB, Bolognese MA, Woodson GC, Moffett AH, et al. Denosumab in postmenopausal women with low bone mineral density. N Engl J Med. 2006;354(8):821–31.
110. Lewiecki EM, Miller PD, McClung MR, Cohen SB, Bolognese MA, Liu Y, et al. Two-year treatment with denosumab (AMG162) in a randomized phase 2 study of postmenopausal women with low BMD. J Bone Miner Res. 2007;22(12):1832–41.
111. Cummings SR, San Martin J, McClung MR, Siris ES, Eastell R, Reid IR, et al. Denosumab for prevention of fractures in postmenopausal women with osteoporosis. N Engl J Med. 2009;361(8):756–65.
112. Beck TJ, Lewiecki EM, Miller PD, Felsenberg D, Liu Y, Ding B, et al. Effects of denosumab on the geometry of the proximal femur in postmenopausal women in comparison with alendronate. J Clin Densitom. 2008;11(3):351–9.
113. Farrier AJ, Sanchez Franco LC, Shoaib A, Gulati V, Johnson N, Uzoigwe CE, et al. New anti-resorptives and antibody mediated anti-resorptive therapy. Bone Joint J. 2016;98-B(2):160–5.
114. Trouvin AP, Goëb V. Receptor activator of nuclear factor-kB ligand and osteoprotegerin: maintaining the balance to prevent bone loss. Clin Interv Aging. 2010;5:345–54.
115. Harslof T, Langdahl BL. New horizons in osteoporosis therapies. Curr Opin Pharmacol. 2016;28:38–42.
116. Hodder A, Huntley C, Aronson JK, Ramachandran M. Pycnodysostosis and the making of an artist. Gene. 2014;119(14):1109–13.
117. Gelb BD, Shi GP, Chapman HA, Desnick RJ. Pycnodysostosis, a lysosomal disease caused by cathepsin K deficiency. Science. 1996;273(5279):1236–8.
118. Mukherjee K, Chattopadhyay N. Pharmacological inhibition of cathepsin K: a promising novel approach for postmenopausal osteoporosis therapy. Biochem Pharmacol. 2016;117:10–9.
119. Jensen AB, Wynne C, Ramirez G, He W, Song Y, Berd Y, et al. The cathepsin K inhibitor odanacatib suppresses bone resorption in women with breast cancer and established bone metastases: results of a 4 week, double-bind, randomized, controlled trial. Clin Breast Cancer. 2010;10(6):452–8.

120. Eastell R, Nagase S, Ohyama M, Small M, Sawyer J, Boonen S, et al. Safety and efficacy of the cathepsin K inhibitor ONO-5334 in postmenopausal osteoporosis: the OCEAN study. J Bone Miner Res. 2012;26(6):1303–12.
121. Zou W, Kitaura H, Reeve J, Long F, Tybulewicz VLJ, Shattil SJ, et al. Syk, c-Src, the αvβ3 integrin, and ITAM immunoreceptors, in concert, regulate osteoclastic bone resorption. J Cell Biol. 2007;176(6):877–88.
122. Clézardin P. Integrins in bone metastasis formation and potential therapeutic implications. Curr Cancer Drug Targets. 2009;9(7):801–6.
123. Murphy MG, Cerchio K, Stoch SA, Gottesdiener K, Wu M, Recker R. Effect of L-0845704, an aVb3 integrin antagonist, on markers of bone turnover and bone mineral density in postmenopausal osteoporotic women. J Clin Endocrinol Metab. 2005;90(4):2022–8.
124. Chambers TJ. Phagocytosis and trypsin-resistant glass adhesion by osteoclasts in culture. J Pathol. 1979;127(2):55–60.
125. Chambers TJ. Resorption of bone by mouse peritoneal macrophages. J Pathol. 1981;135(4):295–9.
126. Zambonin ZA, Teti A, Primavera MV. Isolated osteoclasts in primary culture: first observations on structure and survival in culture media. Anat Embryol (Berl). 1982;165(3):405–13.
127. David JP, Neff L, Chen Y, Rincon M, Horne W, Baron R. A new method to isolate large numbers of rabbit osteoclasts and osteoclast like cells: application to the characterization of serum response element binding proteins during osteoclast differentiation. J Bone Miner Res. 1998;13(11):1730–8.
128. Teti A, Taranta A, Villanova I, Recchia I, Migliaccio S. Osteoclast isolation: new developments and methods. J Bone Miner Res. 1999;14(7):1251–2.
129. Matayoshi A, Brown C, Di Persio JF, Haug J, Abu-Amer Y, Liapis H, et al. Human blood-mobilized hematopoietic precursors differentiate into osteoclasts in the absence of stromal cells. Proc Natl Acad Sci U S A. 1996;93(20):10785–90.
130. Caselli G, Mantovanini M, Gandolfi CA, Allegretti M, Fiorentino S, Pellegrini L, et al. Tartronates: a new generation of drugs affecting bone metabolism. J Bone Miner Res. 1997;12(6):972–81.

Bone Turnover Markers

Katharina Kerschan-Schindl, Ursula Föger-Samwald, and Peter Pietschmann

4.1 Introduction – 57

4.2 Bone Resorption Markers – 58
4.2.1 Hydroxyproline – 58
4.2.2 Pyridinium Cross-Links – 58
4.2.3 Telopeptides of Collagen Type 1 – 58
4.2.4 Tartrate-Resistant Acid Phosphatase Type 5 – 59

4.3 Bone Formation Markers – 59
4.3.1 Propeptides of Type 1 Collagen – 59
4.3.2 Bone-Specific Alkaline Phosphatase – 59
4.3.3 Osteocalcin – 60

4.4 Animal Studies – 60

4.5 Further Bone Turnover Markers – 60
4.5.1 RANK/RANKL/OPG System – 61
4.5.2 Cathepsin K – 61
4.5.3 Periostin – 61
4.5.4 Dickkopf 1 – 61
4.5.5 Sclerostin – 61
4.5.6 Fibroblast Growth Factor 23 – 62
4.5.7 MicroRNAs – 62

4.6 Non-pathological and Pathological Conditions Leading to Changes in BTMs – 62
4.6.1 Bone Turnover Markers in Patients with Chronic Kidney Disease – 62

© Springer International Publishing AG 2017
P. Pietschmann (ed.), *Principles of Bone and Joint Research*, Learning Materials in Biosciences,
DOI 10.1007/978-3-319-58955-8_4

4.7 **Changes in Established Bone Turnover Markers with Osteoporosis Treatment – 63**

4.8 **Clinical Relevance of Bone Turnover Markers – 63**

References – 64

Bone Turnover Markers

What You Will Learn in This Chapter

Bone is a very active tissue that is constantly remodeled in order to adapt to mechanic or metabolic requirements. Since disturbances in bone remodeling may result in relevant skeletal diseases, there is great interest in tools to assess bone remodeling for scientific and clinical applications. Bone turnover markers (BTMs) are among these tools; this chapter presents established markers of bone resorption and bone formation and also selected novel markers. For each marker the biochemical background will be described; in addition you will learn about non-pathological and pathological conditions that lead to alterations of the levels of BTMs. Finally, strengths and weaknesses of the BTMs – especially with regard to their potential clinical applications – will be discussed.

4.1 Introduction

The diagnosis of osteoporosis is based on bone mineral density (BMD) measurements. Dual-energy x-ray absorptiometry (DXA) is believed to be the gold standard method, and therefore, the World Health Organization's (WHO) definition of osteoporosis is based on DXA criteria. Nevertheless, bone quality and thus bone strength are not only determined by BMD but bone architecture (geometry, tissue connectivity, and porosity), material properties, and bone turnover as well. Bone is metabolically active, and approximately 20% of bone tissue is renewed annually in a continuous process of remodeling. Throughout lifetime, remodeling is necessary for different reasons: to regulate calcium homeostasis and to preserve mechanical integrity. Continuous bone turnover is important for adaptation to changing mechanical demands, to remove arising microcracks, and to replace apoptotic cells. In bone remodeling, resorption and formation are tightly coupled and should be balanced. Pathological and age-related imbalances between the amount and activity of osteoclasts (◘ Fig. 4.1) and osteoblasts, however, cause bone loss or disappropriate increases of bone mass. Many different factors regulating bone turnover have been detected so far.

Bone as an endocrine organ expresses such biomarkers, and systemic dispersion makes biochemical measurement of the released factors – the bone turnover markers – possible. However, until now laboratory methods for BTMs are not as well standardized

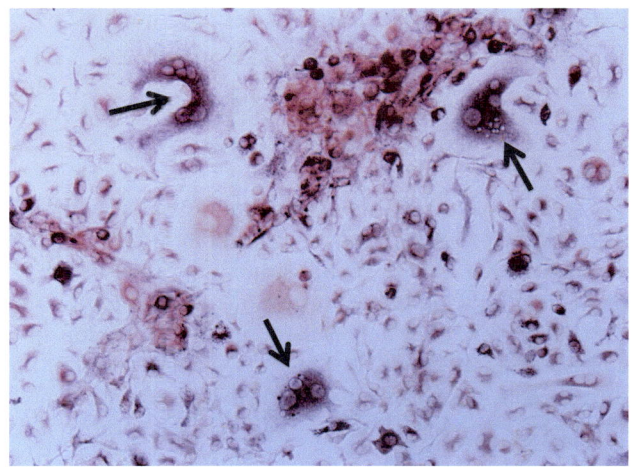

◘ Fig. 4.1 Osteoclasts generated in vivo from murine bone marrow cultures. At the end of the culture period, cells were stained for TRAP by histochemistry. TRAP-positive cells containing three or more nuclei are considered as osteoclasts (marked by *arrows*)

as BMD measurements. Additionally, preclinical conditions like the time of the day of blood withdrawal as well as nutritional status may influence BTMs. These methodological drawbacks are probably the reason why biomarkers were not included in international diagnostic tools like the fracture risk assessment tool (FRAX). In specialized institutions with more experience, standardization seems to be much better and, thus, make data on BTMs more comparable. Hence, BTMs are not only valuable research tools; they also give relevant information on rates of bone resorption and bone formation in the management of osteoporosis patients. For females and males as well, a significant association between BTMs and the risk of future fractures has been shown [1]. In postmenopausal women, prediction of fracture risk by BTMs is actually supposed to be independent from BMD [2].

4.2 Bone Resorption Markers

Most resorption markers (hydroxyproline, pyridinoline, deoxypyridinoline, and crosslaps) are collagen degradation products. A non-collagenous bone resorption marker is the enzyme tartrate-resistant acid phosphatase type 5. Both in research and clinical settings, the most frequently used marker is serum crosslaps.

4.2.1 Hydroxyproline

The first marker of bone resorption was hydroxyproline, an amino acid present in collagen. Since it is not specific for bone resorption and is influenced by the consumption of foods containing collagen, it is not used anymore.

4.2.2 Pyridinium Cross-Links

Breakdown of collagen during bone resorption leads to the release of small fragments with characteristic cross-links – pyridinoline and deoxypyridinoline – into the circulation. As they are excreted by the kidney, they can be measured in urine, which was widely done in the 1990s. However, particularly due to the availability of serum/plasma markers of bone resorption, their clinical relevance has declined.

4.2.3 Telopeptides of Collagen Type 1

The most frequently used bone resorption markers are linear fragments of collagen type 1, which is the major component of the bone's organic matrix. During collagen breakdown, amino-terminal (NTX-1) and carboxy-terminal (CTX-1, also known as crosslaps) telopeptides of type 1 collagen are generated by the enzymatic activity of the cysteine proteinase cathepsin K. They are released into the circulation and can be measured by ELISAs or automated analyzers. NTX-1 and CTX-1 are cleared by the kidneys and, thus, can also be measured in urine. However, such measurements imply the necessity of correction for creatinine excretion. The International Osteoporosis Foundation has recommended to use serum levels of CTX-1 as the reference marker for bone resorption [2, 3].

ICTP is another carboxy-terminal telopeptide of type 1 collagen fragment released during collagen degradation. It can be determined in the peripheral circulation and, in contrast to CTX-1, is supposed to be generated by the enzymatic activity of matrix metalloproteinases [4].

4.2.4 Tartrate-Resistant Acid Phosphatase Type 5

Tartrate-resistant acid phosphatase type 5 (TRAP) is an enzyme; whereas type 5a is produced by macrophages and dendritic cells, type 5b is expressed by osteoclasts and – in consequence – is a marker of bone resorption [5]. TRAP 5b measured in serum does not correspond to the biological activity of osteoclasts but rather to osteoclast number. An advantage of this enzyme is that it is not influenced by renal function; thus, it reflects bone turnover in renal osteodystrophy [6].

4.3 Bone Formation Markers

Within each of the two to five million bone remodeling units, bone resorption only lasts for about 2 weeks and is followed by osteoid production. The responsible cells are osteoblasts, which also express enzymes essential for the mineralization of osteoid. This process takes a few months – a much longer time span than bone resorption. At present the most specific and sensitive markers of bone formation are serum osteocalcin, bone-specific alkaline phosphatase, and the procollagen type 1N-terminal propeptide.

4.3.1 Propeptides of Type 1 Collagen

Around 90% of osteoid is type 1 collagen. The synthesis of type 1 collagen involves various steps including the formation of a procollagen molecule containing C-terminal and N-terminal propeptides (P1CP, P1NP) In one of the last steps leading to mature collagen fibrils, P1CP and P1NP are removed by collagen peptidases and released into the circulation where they can be measured as early bone formation markers [7, 8]. However, since type 1 collagen is also produced in other tissues – for instance, the skin and tendons – P1CP and P1NP are not specific to the bone. Additionally, serum levels of P1NP are dependent on liver function. Nevertheless, the International Osteoporosis Foundation has recommended to use P1NP as the reference marker for bone formation [2, 3].

4.3.2 Bone-Specific Alkaline Phosphatase

Matrix mineralization starts about 2 weeks after osteoid formation. Alkaline phosphatase (AP) is an essential protein involved in this process. Serum levels of total AP, however, show detectable variations only with pronounced changes in bone turnover (e.g., in Paget's disease of bone). Total AP consists of different isoenzymes; a large part of the circulating AP is the liver isoenzyme. Therefore, in clinical routine, not total AP but the bone-specific isoenzyme (BAP) is used to detect minor changes in bone turnover.

4.3.3 Osteocalcin

In mineralized bone the non-collagenous peptide osteocalcin (Oc) is bound to hydroxyapatite. The affinity of Oc to hydroxyapatite depends on its carboxylation which is regulated by vitamin K. This implies that high amounts of uncarboxylated Oc associated with vitamin K deficiency induce a lower affinity of Oc to hydroxyapatite. Oc, which is frequently evaluated in the management of patients with osteoporosis, is supposed to be a late marker of osteoblast development. High levels of osteocalcin have been shown to be associated with an elevated risk of hip fractures in osteoporotic women [9].

4.4 Animal Studies

The structure of essential proteins such as type 1 collagen, the major organic constituent of osteoid, and the regulatory mechanism of central processes such as bone formation and bone resorption are usually highly conserved across different species. Accordingly, the use of BTMs is not restricted to studies in humans, and traditional markers of bone formation (PINP, BAP, Oc) or bone resorption (CTX, TRAP5b) were successfully determined for research purposes in rodents as well as in large animals such as pigs [10–12].

4.5 Further Bone Turnover Markers

Besides the well-established BTMs, more novel biochemical markers of bone metabolism exist. An overview of the regulating factors of bone resorption and bone formation is given in ◘ Fig. 4.2.

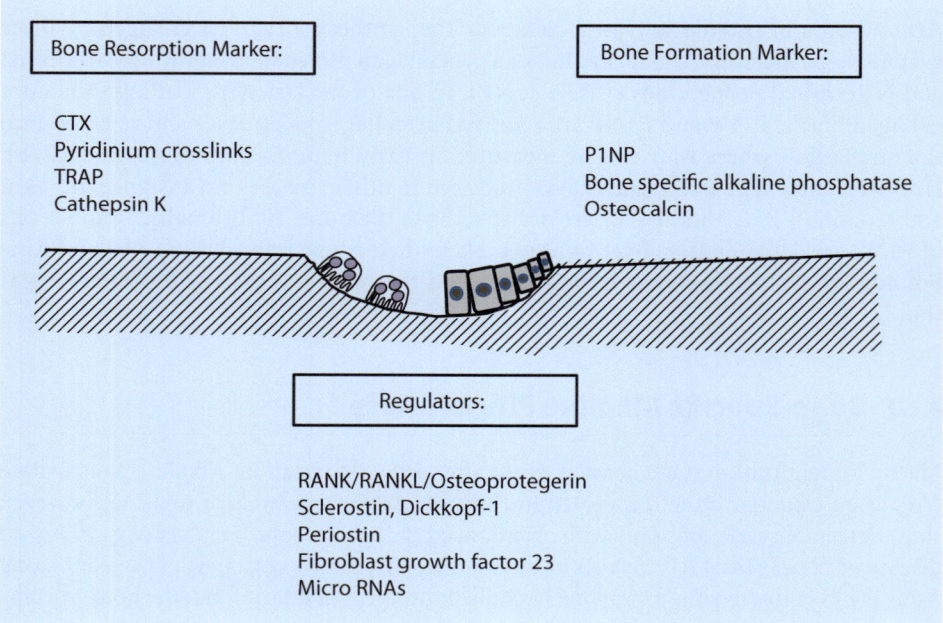

◘ **Fig. 4.2** Schematic representation of markers and regulators of bone turnover

4.5.1 RANK/RANKL/OPG System

Receptor activator nuclear factor κ B (RANK) is a type 1 membrane protein expressed on the surface of osteoclasts (and their precursors) and dendritic cells. Receptor activator nuclear factor κ B ligand (RANKL) is found on the surface of osteoblasts, osteocytes, stromal cells, and T cells. RANK-RANKL interaction in the presence of M-CSF is essential for the formation of osteoclasts. Osteoprotegerin (OPG), which is a member of the tumor necrosis factor (TNF) superfamily, acts as a decoy receptor. By blocking the RANK-RANKL interaction, OPG inhibits the formation of mature osteoclasts. The RANK/RANKL/OPG system is an important factor in coupling of bone resorption and formation. Because of very low serum concentrations, quantification of RANKL levels is sometimes difficult. However, RANKL and OPG concentrations in the peripheral circulation are influenced by age.

4.5.2 Cathepsin K

Cathepsin K (Cat K) is a proteolytic enzyme involved in bone resorption. It is expressed in osteoclasts and released into the resorption lacunae where it catalyzes the degradation of type 1 collagen. Serum Cat K levels are supposed to be increased in patients with osteoporosis, Paget's disease, rheumatoid arthritis, and ankylosing spondylitis (for review, see [13]).

4.5.3 Periostin

Periostin (PSTN) is a protein mainly expressed by periosteal osteoblasts and osteocytes, but it is also released by other collagen-rich tissues such as periodontal ligaments, tendons, heart valves, or skin. PSTN may be regarded as an important regulator of bone formation; it promotes osteoblast function and collagen cross-linking [14]. Accordingly, it is involved in the regulation of bone strength not only in physiologic (following mechanical stress) but in pathological conditions (cancer and bone metastases) as well [15]. An interesting finding is that circulating PSTN levels seem to be associated with an increased risk of fragility fractures which was first shown in the OFELY study for postmenopausal women [16] and later on confirmed for lung transplantation recipients [17].

4.5.4 Dickkopf 1

The regulator dickkopf 1 (Dkk1) is a protein involved in embryogenic and adult bone development. Elevated Dkk1 levels are associated with increased bone resorption and decreased bone formation. It acts via inhibition of the Wingless (Wnt) signaling pathway.

4.5.5 Sclerostin

In contrast to the situation in the embryo, postnatally sclerostin is only expressed in osteocytes. As an inhibitor of the Wnt signaling pathway, it has a negative effect on bone formation. Additionally, it promotes the secretion of RANKL. High serum levels of sclerostin are associated with an increased risk of fragility fractures (for review, see [13, 14]).

4.5.6 Fibroblast Growth Factor 23

Fibroblast growth factor 23 (FGF-23), the negative regulator of phosphate concentration in plasma, is a protein (encoded by the FGF 23 gene) expressed by osteocytes. FGF-23 may also suppress 1-alpha-hydroxylase, reducing its ability to activate vitamin D and subsequently impairing calcium absorption. Only recently, evidence for a role of FGF-23 in regulating bone mineralization has been provided [18]. Data on the association between FGF-23 and the risk of fragility fractures are controversial (for review, see [13]). Determination of FGF-23 may be helpful in patients with severe restriction of renal function.

4.5.7 MicroRNAs

MicroRNAs (miRNAs) induce either repression or cleavage of their target mRNAs. In vitro studies have shown that specific miRNAs are important for osteoclast and osteoblast differentiation and function [19, 20]. Meanwhile human studies have confirmed effects of endothelial miRNAs on osteogenic differentiation [21] and that several miRNAs are differentially expressed following an osteoporotic fracture [22]. Currently the role of miRNAs as novel markers of bone metabolism is under intensive investigation.

4.6 Non-pathological and Pathological Conditions Leading to Changes in BTMs

It is known that BTMs have to be interpreted with caution. Besides preanalytical influences, BTMs change with increasing age. Reference intervals for serum concentrations of P1NP, BAP, and CTX were published for premenopausal and postmenopausal women as well as for men [23]. Even physical activity – dependent on intensity and duration – changes bone metabolism [24–26]. Additionally, BTMs may be altered by diseases or medication such as glucocorticoids. These drugs lead to a reduction in osteoblast formation and function as well as osteocyte function. From the first day of treatment, Oc is reduced, whereas in most cases, the resorption marker CTX remains unchanged. Serum levels of RANKL seem to increase, and OPG levels have been shown to decrease (for review, see [27]).

Inflammation, hyperparathyroidism, hyperthyroidism, and renal failure are supposed to increase BTMs, whereas diabetes, statins, thiazide diuretics, and – as mentioned above – glucocorticoids generally decrease BTMs. After a fracture and during immobilization, BTMs are altered as well [27].

4.6.1 Bone Turnover Markers in Patients with Chronic Kidney Disease

Chronic kidney disease (CKD) is associated with bone disease and an increased risk of fragility fractures. Parathyroid hormone – a classical hormone regulating calcium homeostasis – frequently is increased in patients suffering from CKD. The formation marker BAP and the resorption marker TRAP 5b are not influenced by kidney function and thus may be good surrogates to estimate bone turnover [6, 27].

4.7 Changes in Established Bone Turnover Markers with Osteoporosis Treatment

The aim of any osteoporosis-specific therapy is to prevent fragility fractures. As mentioned in the introduction, BMD is a good surrogate marker to estimate fracture risk. However, since the magnitude of change in BMD in response to treatment is small, it takes at least 1 year till an effect of the medication can be ensured, whereas changes in BTMs may be detected even within a few weeks after the initiation of therapy. In consequence, the use of BTMs should allow the early determination of success or failure of treatment. According to several studies, therapy-associated changes in BTMs seem to be good surrogate markers for the subsequent increase in BMD and the reduction in fracture risk. A least significant change of 30% for serum BTMs has been suggested [28].

Anti-resorptive therapy with alendronate, risedronate, or zoledronic acid leads to a decrease of bone formation and bone resorption markers; denosumab therapy massively suppresses bone resorption (up to 86%; for review, see [29]). After the initiation of an anabolic therapy with parathyroid hormone, bone formation markers, especially P1NP levels, increase rapidly. Bone resorption markers increase at a later time point and to a lesser degree opening the so-called anabolic window (for review, see [29]). In the case of anti-resorptive therapy, a more pronounced decrease of BTMs is predictive of a more distinct effect on BMD. In the case of anabolic therapy, a greater increase in serum P1NP is associated with a higher expected effect on BMD [27].

Cessation of osteoporosis-specific therapy changes bone turnover. However, after alendronate treatment, BTMs remain suppressed for several years [30]. In contrast, after discontinuation of denosumab, suppressed BTMs increase more or less immediately [31]. Teriparatide-induced increases of BTMs gradually decline after cessation of therapy [32].

4.8 Clinical Relevance of Bone Turnover Markers

BTMs have limited value in predicting bone loss and fracture risk in the individual subject, but it is known that patients with accelerated bone turnover lose bone at a fast rate. These patients may be even better candidates for an anti-resorptive therapy than those with normal or low turnover. The combination of BMD and BTM assessment after the change of osteoporosis-specific therapy seems to provide a better assessment of the antifracture effect than either measurement alone [33]. Thus, BTMs may be helpful in deciding what to do in special cases. In the following paragraphs, two examples are given.

An osteoporotic woman on oral bisphosphonate therapy is reevaluated after a few years of treatment. BMD is more or less unchanged. In case of low BTMs, which indicate that the patient takes her medication regularly and thus at least partially should respond to treatment, bisphosphonate therapy may be continued. In case of high BTMs, which may be interpreted as noncompliance of the patient or ineffectiveness of the oral bisphosphonate, a switch of the therapeutic regimen would be recommended [34].

The second case also illustrates how BTMs may guide clinical decisions: A frail elderly woman, who is on drug holiday after 5 years of osteoporosis-specific medication, is reevaluated. Her BMD has slightly decreased to an osteoporotic value; the bone resorption marker is elevated. In this case a resumption of osteoporosis-specific therapy is recommended. The reason is threefold: The woman's bone resorption is increased, her BMD has decreased during the drug holiday, and last but not least, she has a high risk of falling [34].

These cases show that determination of bone turnover status in specific cases is helpful in guiding the selection of therapy. The use of BTMs in individual cases is recommended by the German guideline for prevention, diagnosis, and therapy of osteoporosis (► http://www.dv-osteologie.org/uploads/Leitlinie%202014/DVO-Leitlinie%20Osteoporose%202014%20Kurzfassung%20und%20Langfassung%20Version%201a%2012%2001%202016.pdf).

> **Take-Home Messages**
>
> — According to the WHO definition, BMD measurements determine the diagnosis of osteoporosis. Nevertheless, together with clinical and imaging techniques, biochemical tests reflecting bone resorption and bone formation play an important role in the assessment as well as the differential diagnosis of metabolic bone disease. Currently BTMs are not included in the FRAX algorithm, but the German guideline for prevention, diagnosis, and therapy of osteoporosis recommends measurements in individual cases.
> — CTX and P1NP are the most advised markers for evaluation of bone resorption and bone formation.
> — BTMs reflect the efficiency of the therapy and can be used to improve patients' compliance, a very important aspect in reaching our treatment goal.

References

1. Johansson H, Oden A, Kanis JA, et al. A meta-analysis of reference markers of bone turnover for prediction of fracture. Calcif Tissue Int. 2014;94:560–7.
2. Vasikaran S, et al. Markers of bone turnover for the prediction of fracture risk and monitoring of osteoporosis treatment: a need for international reference standards. Osteoporos Int. 2011;22:391–420.
3. Vasikaran S, et al. International Osteoporosis Foundation and International Federation of Clinical Chemistry and Laboratory medicine. Position on bone markers standards in osteoporosis. Clin Chem Lab Med. 2011;49:1271–4.
4. Garnero P, Ferreras M, Karsdal MA, Nicamhlaoibh R, Risteli J, Borel O, Qvist P, Delmas PD, Foged NT, Delaissè JM. The type I collagen fragments ICTP and CTX reveal distinct enzymatic pathways of bone collagen degradation. J Bone Miner Res. 2003;18:859–67.
5. Halleen JM, Tiitinen SL, Ylipahkala H, Fagerlund KM, Vaananen HK. Tartrate-resistant acid phosphatase 5b (TRACP 5b) as a marker of bone resorption. Clin Lab. 2006;52:499–509.
6. Fahrleitner-Pammer A, Herberth J, Browning SR, Obermayer-Pietsch B, Wirnsberger G, Holzer H, Dobnig H, Malluche HH. Bone markers predict cardiovascular events in chronic kidney disease. J Bone Miner Res. 2008;23:1850–8.
7. Melkko J, Niemi S, Risteli L, Risteli J. Radioimmunoassay for the carboxyterminal propeptide of human type I procollagen (PICP). Clin Chem. 1990;36:1328–32.
8. Melkko J, et al. Immunoassay for intact amino-terminal propeptide of human type I procollagen. Clin Chem. 1996;42:947–54.
9. Vergnaud P, et al. Undercarboxylated osteocalcin measured with a specific immunoassay predicts hip fracture in elderly women; the EPIDOS study. J Clin Endocrinol Metab. 1997;82:719–24.
10. Pietschmann P, Skalicky M, Kneissel M, Rauner M, Hofbauer G, Stupphann D, Viidik A. Bone structure and metabolism in a rodent model of male senile osteoporosis. Exp Gerontol. 2007;42:1099–108.
11. Sipos W, Zysset P, Kostenuik P, Mayrhofer E, Bogdan C, Rauner M, Stolina M, Dwyer D, Sommerfeld-Stur I, Pendl G, Resch H, Dall'Ara E, Varga P, Pietschmann P. OPG-Fc treatment in growing pigs leads to rapid reduction in bone resorption markers, serum calcium, and bone formation markers. Horm Metab Res. 2011;43:944–9.

12. Rauner M, Föger-Samwald U, Kurz MF, Brünner-Kubath C, Schamall D, Kapfenberger A, Varga P, Kudlacek S, Wutzl A, Höger H, Zysset PK, Sh GP, Hofbauer LC, Sipos W, Pietschmann P. Cathepsin S controls adipocytic and osteoblastic differentiation, bone turnover, and bone microarchitecture. Bone. 2014;64:281–7.
13. Garnero P. New developments in biological markers of bone metabolism. Bone. 2014;66:46–55.
14. Bonnet N, Garnero P, Ferrari S. Periostin action in bone. Mol Cell Endocrinol. 2016;432:75–82.
15. Merle B, Garnero P. The multiple facets of periostin in bone metabolism. Osteoporos Int. 2012;23:1199–212.
16. Rousseau JC, Sornay-Rendu E, Bertholon C, Chapurlat R, Garnero P. Serum periostin is associated with fracture risk in postmenopausal women: a 7 years prospective analysis of the OFELY study. J Clin Endocrinol Metab. 2014;99:2533–9.
17. Kerschan-Schindl K, Ebenbichler G, Gruther W, Petrikic A, Föger-Samwald U, Kudlacek S, Patsch J, Jaksch P, Klepetko W, Pietschmann P. Serum levels of musculoskeletal markers of lung transplantation recipients (manuscript in preparation).
18. Murali SK, Roschger P, Zeitz U, Klaushofer K, Andrukhova O, Erben RG. FGF23 regulates bone mineralization in a 1,25(OH)2 D3 and Klotho-independent manner. J Bone Miner Res. 2016;31(1):129–42.
19. Lian JB, Stein HS, van Wijnen AJ, Stein JL, Hassan MQ, Gaur T, et al. Micro RNA control of bone formation and homeostasis. Nat Rev Endocrinol. 2012;8:212–7.
20. Sugatani T, Hruska KA. Impaired micro-RNA pathways diminish osteoclast differentiation and function. J Biol Chem. 2009;284:4667–78.
21. Weilner S, Schraml E, Wieser M, Messner P, Schneider K, Wasermann K, Micutkova L, Fortschegger K, Maier AB, Westendorp R, Resch H, Wolbank S, Redl H, Jamsen-Dürr P, Pietschmann P, Grillari-Voglauer R, Grillari J. Secreted microvascular miR-31 inhibits osteogenic differentiation of mesenchymal stem cells. Aging Cell. 2016;15:744–54.
22. Weilner S, Skalicky S, Salzer B, Keider V, Wagner M, Hildner F, Gabriel C, Dovjak P, Pietschmann P, Grillari-Voglauer R, Grillari J, Hackl M. Differentially circulating miRNAs after recent osteoporotic fractures can influence osteogenic differentiation. Bone. 2015;79:43–51.
23. Michelsen J, Wallaschofski H, Friedrich N, Spelhagen C, Rettig R, Ittermann T, Nauck M, Hannemann A. Reference intervals for serum concentrations of three bone turnover markers for men and women. Bone. 2013;57:399–404.
24. Kerschan-Schindl K, Thalmann M, Sodeck GH, Skenderi K, Matalas AL, Grampp S, Ebner C, Pietschmann P. A 246-km continuous running race causes significant changes in bone metabolism. Bone. 2009;45:1079–83.
25. Kerschan-Schindl K, Thalmann M, Weiss E, Tsironi M, Föger-Samwald U, Meinhart J, Skenderi K, Pietschmann P. Changes in serum levels of myokines and Wnt-antagonists after and ultramarathon race. PLoS One. 2015;10:e0132478.
26. Lombardi G, Sanchis-Gomar F, Perego S, Sansoni V, Banfi G. Implications of exercise-induced adipomyokines in bone metabolism. Endocrine. 2016;54:284–305.
27. Cavalier E, Bergmann P, Bruyere O, Delanaye P, Durnez A, Devogelaer JP, Ferrari SL, Gielen E, Goemare S, Kaufmann JM, Nzeusseu Toukap A, Feginster JY, Rousseau AF, Rozenberg S, Scheen AJ, Body JJ. The role of biochemical of bone turnover markers in osteoporosis and metabolic bone disease: a consensus paper of the Belgian Bone Club. Osteoporos Int. 2016;27:2181–95.
28. Bergmann P, Body JJ, Boonen S, et al. Evidence-based guidelines for the use of biochemical markers of bone turnover in the selection and monitoring of bisphosphonate treatment in osteoporosis: a consensus document of the Belgian Bone Club. Int J Clin Pract. 2009;63:19–26.
29. Bandeira F, Costa AG, Filho MAS, Pimentel L, Lima L, Bilezikian JP. Bone markers and osteoporosis therapy. Arq Bras Endocrinol Metab. 2014;58:504–13.
30. Ensrud KE, Barrett-Connor EL, Schwartz A, Santora AC, Bauer DC, Suryawanshi S, Feldstein A, Haskell WL, Hochberg MC, Torner JC, Lombardi A, Black DM, Fracture Intervention Trial Long-Term Extension Research Group. Randomized trial of effect of alendronate continuation versus discontinuation in women with low BMD: results from the Fracture Intervention Trial long-term extension. J Bone Miner Res. 2004;19:1259–69.
31. Miller PD, Bolognese MA, Lewiecki EM, McClung MR, Ding B, Austin M, Liu Y, San Martin J, Amg Bone Loss Study Group. Effect of denosumab on bone density and turnover in postmenopausal women with low bone mass after long-term continued, discontinued, and restarting of therapy: a randomized blinded phase 2 clinical trial. Bone. 2008;43:222–9.

32. Glover S, Eastell R, McCloskey EV, Rogers A, Garnero P, Lowery J, Belleli R, Wright TM, John MR. Rapid and robust response of biochemical markers of bone formation to teriparatide therapy. Bone. 2009;45:1053–8.
33. Bouxsein ML, Delmas PD. Considerations for development of surrogate endpoints for antifracture efficacy of new treatments in osteoporosis: a perspective. J Bone Miner Res. 2008;23:1155–67.
34. Kerschan-Schindl K, Mikosch P, Obermayer-Pietsch B, Gasser RW, Dimai HP, Fahrleitner-Pammer A, Dobnig H, Roschger P, Preisinger E, Klaushofer K, Resch H, Pietschmann P. Current controversies in clinical management of postmenopausal osteoporosis. Exp Clin Endocrinol Diabetes. 2014;122:437–44.

Bone Imaging by Nuclear Medicine

Peter Mikosch

5.1	Introduction – 68	
5.2	Bone Scintigraphy: General Information – 68	
5.3	Bone Disorders with Pathological Bone Scintigraphies – 70	
5.3.1	Primary Bone Tumors – 70	
5.3.2	Bone Metastases – 72	
5.4	Metabolic Bone Disorders – 74	
5.4.1	Osteoporosis – 74	
5.4.2	Osteomalacia – 75	
5.4.3	Primary Hyperparathyroidism – 75	
5.4.4	Renal Osteodystrophy – 77	
5.4.5	Hyperthyroidism – 77	
5.4.6	Paget's Disease of Bone – 77	
5.4.7	Fibrous Dysplasia – 77	
5.5	Inflammation – 78	
5.5.1	Joint Inflammation – 78	
5.5.2	Chronic Osteomyelitis – 78	
5.5.3	Degenerative Bone and Joint Changes – 79	
5.5.4	Algodystrophy – 80	
5.6	Conclusion – 82	
	References – 82	

© Springer International Publishing AG 2017
P. Pietschmann (ed.), *Principles of Bone and Joint Research*, Learning Materials in Biosciences,
DOI 10.1007/978-3-319-58955-8_5

What You Will Learn in This Chapter

The imaging targets of bone scintigraphy are all changes of bone metabolism independent of its underlying pathology. The technical background of this imaging method will be explained briefly. The use of bone scintigraphy in connection with some selected but frequent clinical questions and entities will be outlined concerning the results of bone scintigraphy in each entity. In addition, limitations of bone scintigraphy with technetium-99m (Tc-99m)-labeled diphosphonates and a comparison to other imaging modalities will be presented.

5.1 Introduction

The skeleton including bone and its joints can be affected by a broad variety of disorders ranging from inflammation, benign and malignant bone tumors, bone metastases from other organs to disorders diffusely affecting bone due to changes of bone metabolism. Examples of such disorders affecting bone metabolism are osteoporosis, osteomalacia, renal osteodystrophy, or hyperthyroidism. The diagnostic approach to these bone pathologies starts with the medical history and physical examination, as in general by other organ systems, as well. However, a closer insight by physical examination into bone pathology cannot be done easily. Bone biopsy may be necessary in some circumstances to obtain information whether, e.g., an unclear pathological bone lesion is of benign or malignant origin, an information essential ahead in case of possible surgery. However, biopsies of bone are an invasive and time-consuming procedure, not to mention possible complications such as pain, bleeding, or infection. Thus, in most cases imaging results will be the easiest way to access an insight view of bone and in order to set up a diagnosis, as the different disorders reveal typical pathological changes by imaging. Various imaging modalities of bone are available in daily routine (conventional X-ray, computed tomography, magnetic resonance imaging, osteodensitometry, bone scintigraphy). Whereas conventional X-ray, computed tomography, and magnetic resonance imaging reveal information about the morphological aspects of bone and bony disorders, bone scintigraphy gives an insight into the metabolic changes associated with the different disorders. In this respect bone scintigraphy has a particular position in imaging, which stands not in competition to the other morphologically orientated bone imaging modalities; it can rather be seen as a method adding additional information.

The article will give a short introduction into the concepts of bone imaging by bone scintigraphy. Detailed information on bone scintigraphy including further indications, scintigraphic results, and diagnostic limitations can be found in specific literature [1–10].

5.2 Bone Scintigraphy: General Information

Radioisotopic bone scanning using technetium-99m-labeled phosphate and diphosphonate compounds, commonly called bone scintigraphy or bone scan, is a method of imaging metabolic changes of the skeleton. The nuclear medicine technique is sensitive to areas of unusual bone rebuilding activity, since the radiopharmaceutical is taken up by osteoblast cells which build bone. For a bone scan, the patient is injected with a small amount of a diphosphonate (most commonly methylene diphosphonate (MDP) is used), which is labeled with technetium-99m (Tc-99m) as a

Bone Imaging by Nuclear Medicine

gamma-emitting radioactive material for tracer detection by a gamma camera (◘ Table 5.1). Diphosphonates can exchange places with bone phosphate in regions of active bone growth or bone metabolism, thus anchoring the radioisotope to that specific region. The patient can be scanned at different times, e.g., rapid imaging sequences just after injection of the radiopharmaceutical resembling the arterial perfusion phase, after a few minutes to visualize the extent of soft tissue perfusion, and after 2–4 h to image the bone phase (◘ Figs. 5.1 and 5.2). To view small lesions (less

◘ **Table 5.1** Tc-99m-labeled diphosphonates

γ-Emitting tracer: technetium-99m (Tc-99m)

Bone-seeking component

  HMDP: hydroxymethylene diphosphonate

  HEDP: hydroxyethylidene diphosphonate

  HDP: hydroxymethane diphosphonate

  MDP: methylene diphosphonate

Accumulation: in bone after 2–4 h 40%

Excretion: renal excretion after 2–4 h by 50% of injected activity

=> Radiation burden of the bladder: advice to drink much to keep radiation burden as low as possible by increased diuresis

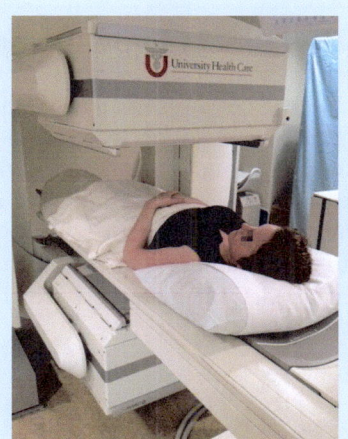

555–740 MBq Tc-99m-Diphosphonat i.v.

- **Perfusion phase/arterial phase**
  directly after tracer application
  1 Frame/sec for 120 sec
- **Bloodpool phase/soft tissue phase**
  after 3–5 min
  static images
- **Delayed phase/bone phase**
  after 3–4 h
  static images

Image interpretation
Visually
ROI technique: calculation of ratios between lesions and normal bone

◘ **Fig. 5.1** Bone scintigraphy – phases of investigation

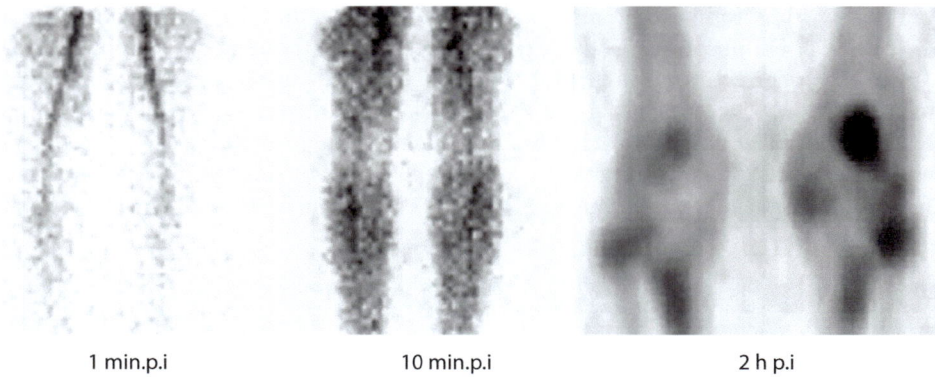

| 1 min.p.i | 10 min.p.i | 2 h p.i |

Fig. 5.2 Perfusion phase/blood pool phase/delayed phase

than 1 cm) – especially in the spine – the SPECT imaging technique may be required. The method has assumed major clinical importance to some bone pathologies, which will be outlined in the following section concerning the scintigraphic results. Other information in connection with these pathologies, such as epidemiology, pathology, radiological results, and course of disease, will be outlined only briefly in this chapter, if necessary for the understanding of the scintigraphic results. For further and broader information on these disorders, additional literature will have to be consulted.

5.3 Bone Disorders with Pathological Bone Scintigraphies

5.3.1 Primary Bone Tumors

Some but not all primary bone tumors, whether benign or malignant, show a pathological accumulation in bone scintigraphy. Bone scintigraphy can thus give information concerning the metabolic activity of the tumor and about whether or not additional pathological lesions are present in the skeleton. However, bone scintigraphy cannot give a conclusive information about the dignity of the tumor, as both benign and malignant bone tumors can show pathologically increased tracer accumulation (Figs. 5.3 and 5.4). A list of tumors which reveals increased tracer accumulation is listed in Table 5.2.

Fig. 5.3 Example of a benign bone tumor – osteoid osteoma

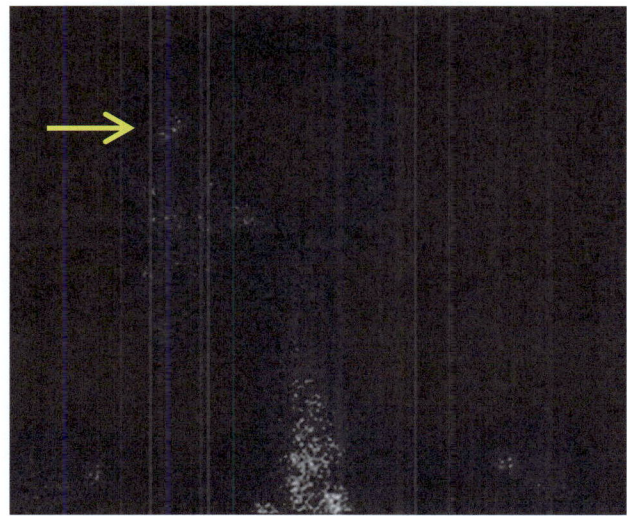

Fig. 5.4 Example of a malignant bone tumor – osteosarcoma

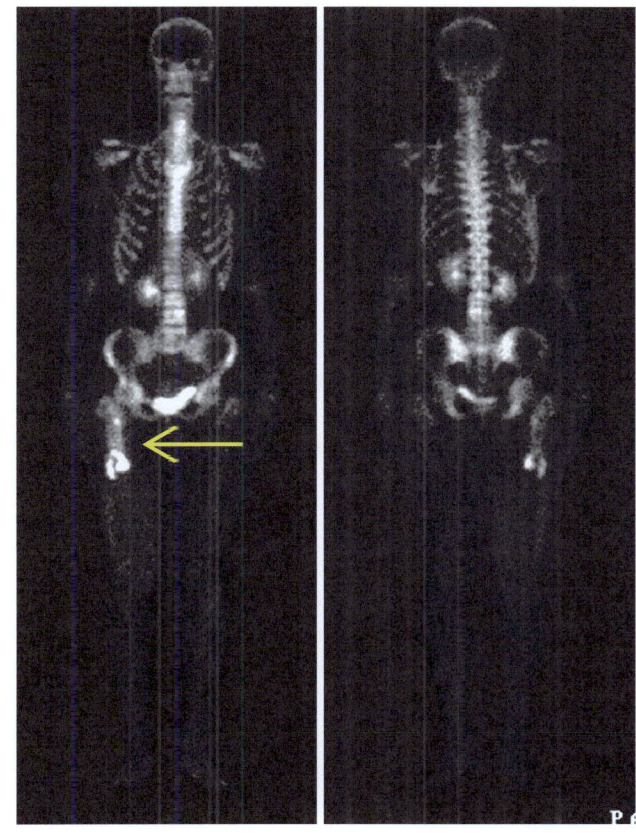

☐ **Table 5.2** Benign and malignant bone tumors with scintigraphic abnormalities

Primary bone tumors
Osteogenic tumors
Osteoid osteoma
Osteoblastoma
Osteosarcoma
Chondrogenic tumors
Chondroma, enchondroma, periosteal chondroma
Osteochondroma
Chondrosarcoma
Primary chondrosarcoma
Secondary chondrosarcoma
Collagenic tumors
Fibrosarcoma
Myelogenic tumors
Ewing sarcoma
Bone metastases
Osteoblastic metastases
Osteolytic metastases

5.3.2 Bone Metastases

Different tumors such as breast cancer, prostate cancer, or lung cancer frequently develop bone metastases. Bone metastases with osteoblastic activity in bone scintigraphy show an increased tracer accumulation (☐ Fig. 5.5), whereas osteoclastic bone metastases in general reveal no scintigraphic abnormalities, as the tracer accumulation is mediated via the osteoblasts but not the osteoclasts. The exception will be large osteoclastic metastases which impose as «cold lesions» (☐ Fig. 5.6). In cases of diffusely spread bone metastases, the skeleton imposes with an increased tracer accumulation. As the complete tracer is bound and trapped in bone, the kidneys by which bone scintigraphic tracers are normally excreted within hours show no tracer accumulation. This scintigraphic appearance is called hyperscan (☐ Fig. 5.7). Although frequently seen in metastatic bone disease, the image of a hyperscan is not specific for metastatic bone disease, as it can also appear in cases of hypermetabolic bone disorders such as hyperthyroidism or hyperparathyroidism (see below).

Bone Imaging by Nuclear Medicine

Fig. 5.5 Multiple bone metastases in prostate cancer

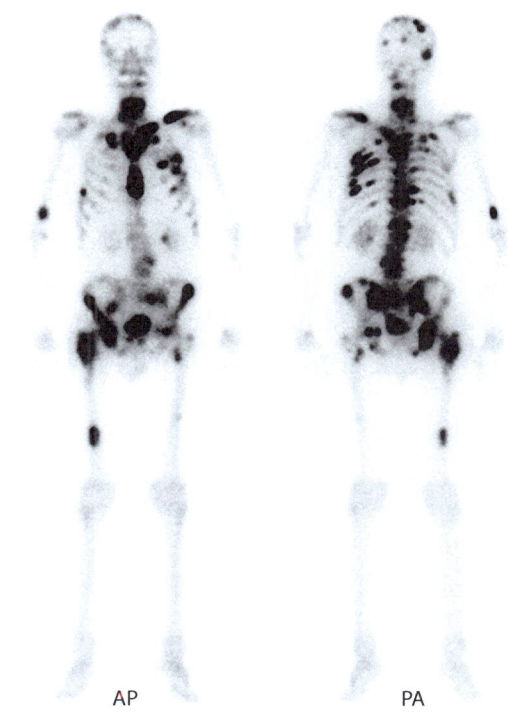

AP PA

Fig. 5.6 Osteolytic bone metastasis in lung cancer

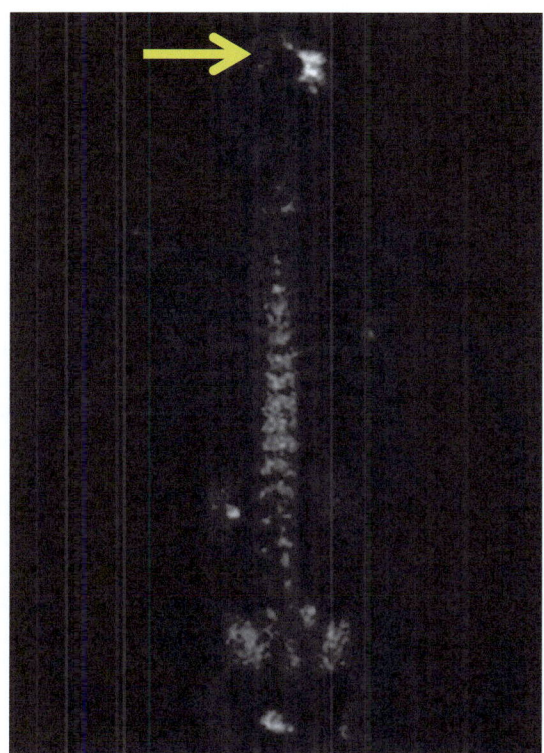

Fig. 5.7 «Hyperscan» – diffuse bone metastases from gastric cancer

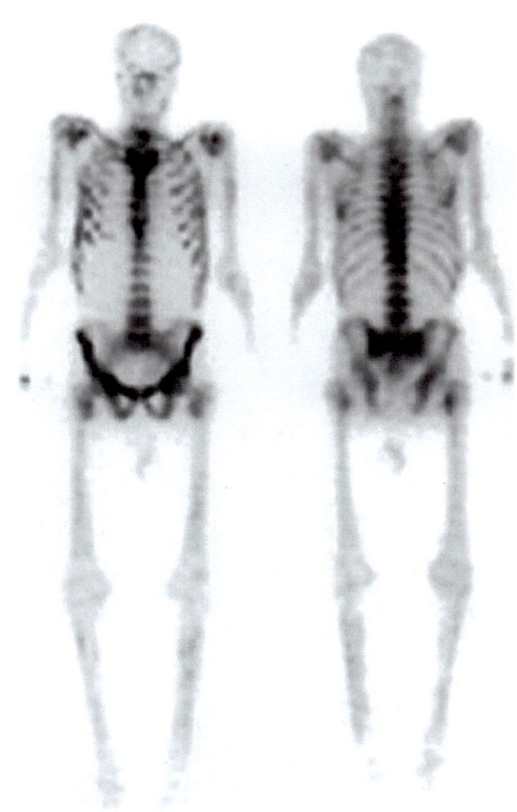

5.4 Metabolic Bone Disorders

5.4.1 Osteoporosis

The decrease of bone mass and the reduction of bone strength, as by which osteoporosis is defined, cannot be visualized by bone scintigraphy. The reduction of bone mass can be measured by dual X-ray absorptiometry (DXA), and the morphological changes of bone including fractures can be imaged by different radiological methods (e.g., X-ray, computed tomography, high-resolution computed tomography). However, bone scintigraphy can be of clinical use in osteoporosis, as not all fractures can be visualized by radiology (occult fractures). After falls non-dislocated fractures are often overseen with conventional radiological methods such as X-rays, although the patients report on persistent pain or difficulties to walk. With bone scintigraphy increased tracer accumulation occurs due to bone hypermetabolism in the area of the fracture. The typical scintigraphic pattern seen in association with fractures is a linear hypermetabolic lesion (◘ Fig. 5.8).

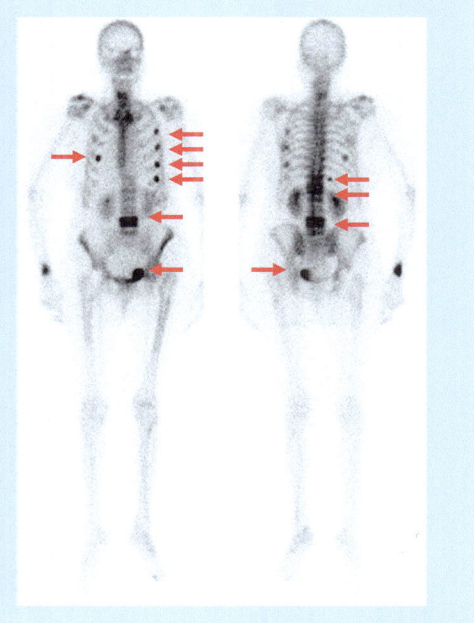

▣ **Fig. 5.8** Multiple osteoporotic fractures after fall

Female, 82a
hip-TEP right side 1992
hip-Fx left side 1995, DHS left hip
fall 3 weeks ago, since that time
pain within the lumbar spine and
chest

Dept. Traumatology: no new Fx

99mMDP bone scintigraphy:
multiple Fx of several ribs,
ramus sup. ossis pubis sin.
T11, T12, L1, L2, L4

5.4.2 Osteomalacia

Although bone scintigraphy is not the method of choice to diagnose osteomalacia, the disease reveals some signs in bone scintigraphy. In severe disease bones show deformities such as bending of long bones, and an accentuation of tracer accumulation in the cortical region of long bones can frequently be visualized (▣ Fig. 5.9).

5.4.3 Primary Hyperparathyroidism

Bone turnover is accelerated in hyperparathyroidism which can be seen in bone scintigraphy as an increased tracer accumulation in particular in the skull and the central skeleton accompanied with a decreased or missing visualization of the kidneys (hyperscan) (▣ Fig. 5.10).

In severe disease brown tumors can occur which usually show an increased tracer accumulation with bone scintigraphy. This can potentially be misled as bone tumors or bone metastases.

Female, 86yrs
Resident of nursing home
25-OH Vit D3 15nmol/l

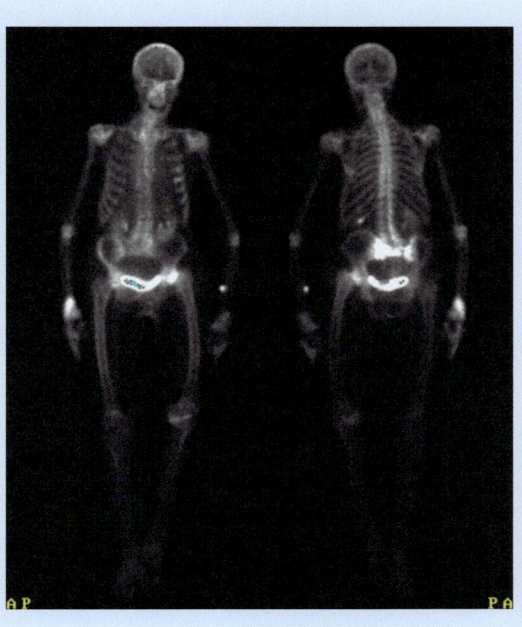

Fig. 5.9 Osteomalacia

Female, 73yrs,
papillary thyroid carcinoma
(surgery + iodine TE 1971)
Tg < 0.5ng/ml R101% - remission

X-ray: osteoporosis

Bone scintigraphy

diffusely increased tracer upake
=> metabolic osteopathy

Laboratory parameters:
fT4	22.49	Ca	2.46
TT3	1.76	Ca++	1.23
TSH	<0.03	PTH	96.4 pg/ml
osteocalcin			64.8 ng/ml
serum-Crosslaps			750.4 pmol/l
25-OH Vit. D			17.9 ng/ml

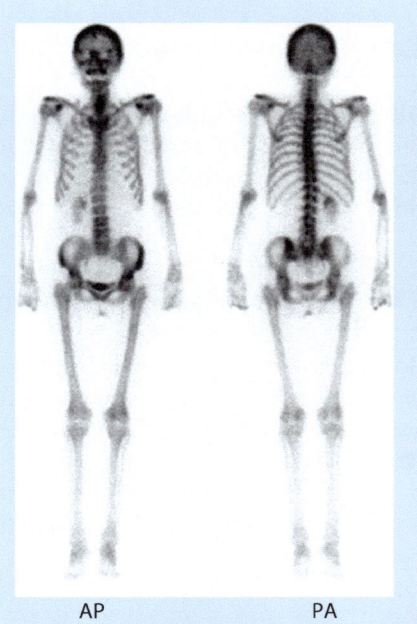

Fig. 5.10 Hypermetabolic bone disease

5.4.4 Renal Osteodystrophy

Tracer accumulation can be increased or decreased, and no typical scintigraphic features are associated with renal osteodystrophy. Usually also scintigraphic features as seen with osteomalacia, osteoporosis, or primary hyperparathyroidism can be found in patients with renal osteodystrophy. In general scintigraphy provides no relevant additional information in this metabolic bone disease.

5.4.5 Hyperthyroidism

In patients with clinical hyperthyroidism, bone turnover is extensively increased with an acceleration of bone formation and bone resorption. This increased bone turnover can be visualized by bone scintigraphy revealing a diffusely increased tracer accumulation in the whole skeleton, thus producing the image of a hyperscan (increased tracer accumulation in the whole skeleton but no visualization of the kidneys as no tracer is excreted) (◘ Fig. 5.10). This scintigraphic pattern can be also seen in patients with diffuse bone metastases (◘ Fig. 5.7).

5.4.6 Paget's Disease of Bone

Paget's disease of bone is a chronic disorder that can result in enlarged and misshapen bones caused by the excessive breakdown and formation of bone, followed by disorganized bone remodeling. This results in weak and misshapen bone prone to fracture, pain, and arthritis. Rarely, it can develop into a primary bone cancer known as Paget's sarcoma. Often Paget's disease is localized to one or only a few bones in the body. The skull, pelvis, femur, and lower lumbar vertebrae are the most commonly affected bones.

Paget's disease can occur as monostotic or polyostotic and usually reveals an increased tracer accumulation within the affected bone (◘ Fig. 5.11).

Bone scintigraphy wherein the whole skeleton is scanned with one investigation is most useful in determining the extent and activity of the condition. If a bone scan suggests Paget's disease, the affected bone(s) should be X-rayed to confirm the diagnosis.

5.4.7 Fibrous Dysplasia

Similar to Paget's disease, also lesions of fibrous dysplasia can show an increased tracer accumulation. Although lesions of Paget's disease usually show a more intense tracer accumulation in bone scintigraphy than fibrous dysplasia lesions, a clear differentiation between fibrous dysplasia and Paget's disease by scintigraphic appearance is not possible. The differentiation can rather be made by means of localization and by other radiological methods such as X-ray or computed tomography. The major indication and value of bone scintigraphy is therefore similar to Paget's disease, the scanning of the whole skeleton within one investigation for lesions of this hypermetabolic bone disorder.

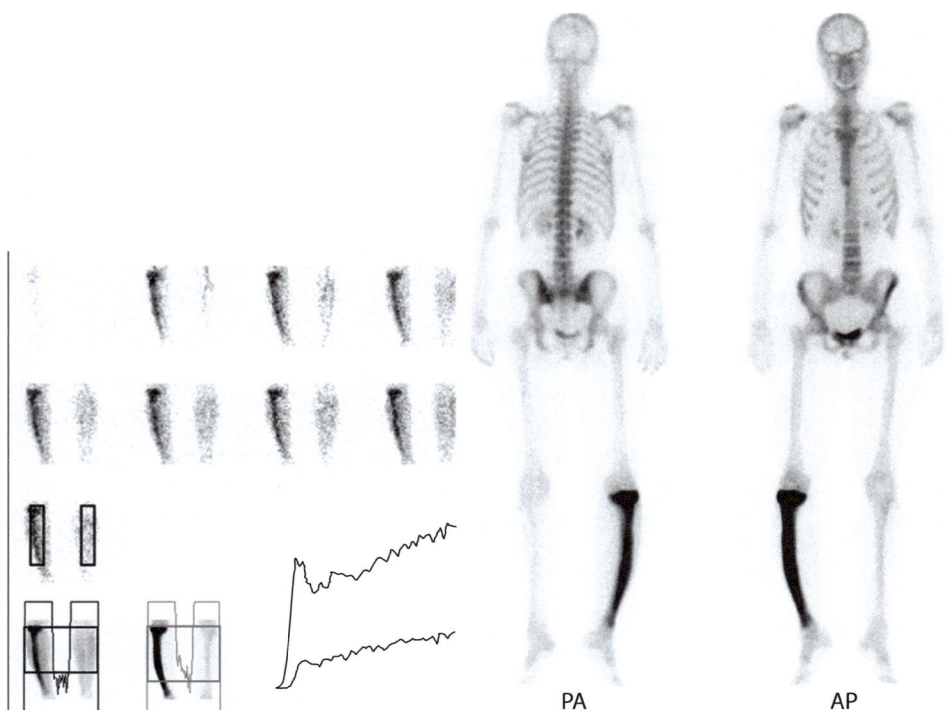

◘ Fig. 5.11 Monostotic Paget's disease

5.5 Inflammation

5.5.1 Joint Inflammation

For the purpose of imaging joint inflammation, bone scintigraphy is usually performed in a three-phase technique with a rapid perfusion/arterial phase, delayed perfusion phase resembling soft tissue perfusion, and a late bone phase (see above). Tracer accumulation is usually not only described by optical appearance but also by a region of interest technique (ROI) with comparison of the affected joint (joint region with the similar contralateral region. In acute joint inflammation, all three phases reveal an increased tracer accumulation. In chronic joint inflammation, the arterial phase is usually normal, but the soft tissue and bone phase show pathological tracer accumulation around the affected joint (◘ Fig. 5.12).

5.5.2 Chronic Osteomyelitis

After complex fractures or orthopedic interventions, chronic osteomyelitis may develop. Bone scintigraphy can show the extent of inflammation within the bone cavity, the intensity of inflammation (◘ Fig. 5.13), or after antibiotic therapy the disappearance of inflammatory activity.

Bone Imaging by Nuclear Medicine

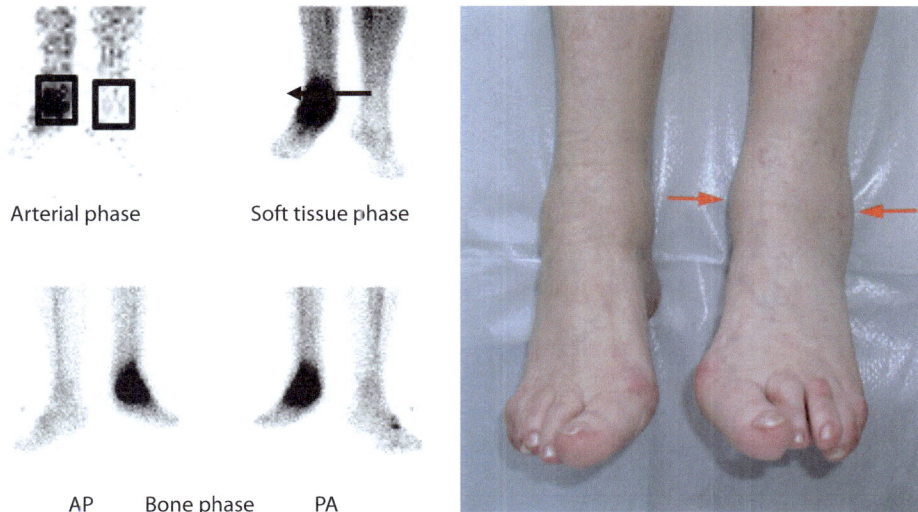

Fig. 5.12 Acute joint inflammation – acute arthritis

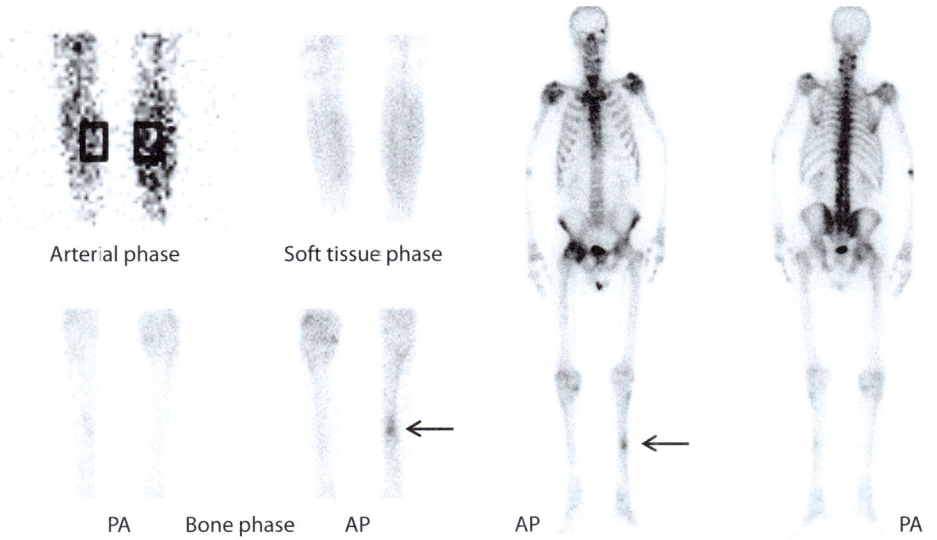

Fig. 5.13 Chronic osteomyelitis

5.5.3 Degenerative Bone and Joint Changes

In particular within the spine or the large joints, an increased tracer accumulation can frequently be seen due to morphological changes of the joints (arthrosis) or pathological mechanical alterations of skeletal regions (e.g., scoliosis of the spine) (◘ Fig. 5.14). In general the increase of tracer accumulation in association with

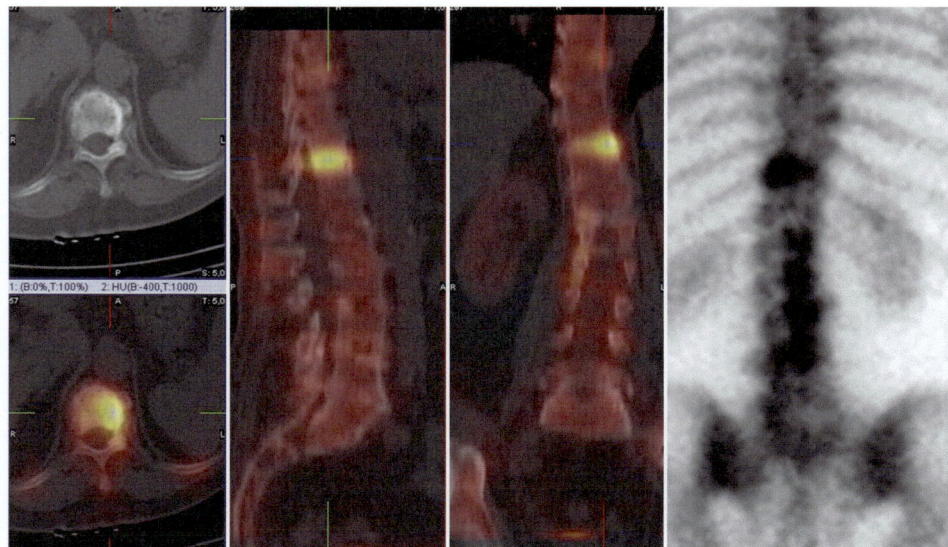

Fig. 5.14 Degenerative bone changes – osteochondrosis

degenerative bone disease is mild to moderate, so scintigraphically in a differentiation to other focal bone pathologies such as inflammation or bone tumors, bone metastasis is possible.

5.5.4 Algodystrophy

Persistent pain or atrophy may occur after fracture within the fractured region or after surgery. Bone scintigraphy performed as three-phase scintigraphy reveals pathologic patterns in algodystrophy within the whole affected extremity, which permits the establishment of the diagnosis and extent of disturbance (Fig. 5.15). The different patterns dependent on the time since the onset of symptoms are listed in Table 5.3.

Bone Imaging by Nuclear Medicine

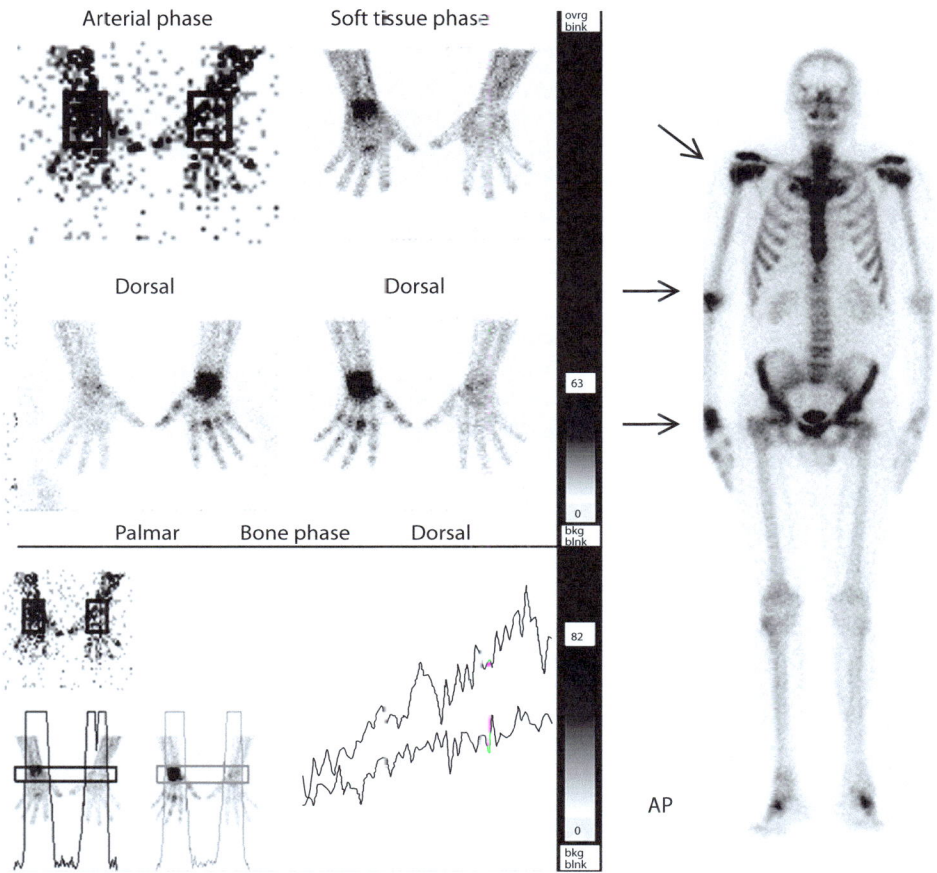

Fig. 5.15 Reflex sympathetic dystrophy (RSD)

Table 5.3 Algodystrophy, reflex sympathetic dystrophy (RSD)

	Perfusion	Blood pool	Late bone scan
Early phase	↑	↑	↔
Later phase	↔ ↓	↔ ↓	↑
Chronic phase	↓	↓	↓

5.6 Conclusion

Bone scintigraphy is a method which can easily be performed allowing to scan the whole skeleton on a metabolic level or to visualize pathological changes of defined regions including its perfusion and soft tissue accumulation. Various bone pathologies as outlined above show pathological changes. The strength of bone scintigraphy is a high sensitivity to detect pathological changes in bone metabolism; however, specificity is low. The differentiation between different pathologies can be established through different anatomical patterns of the pathological lesions seen, in combination with arterial perfusion and soft tissue accumulation adjacent to bone pathologies and to some extent by the intensity of tracer accumulation. Furthermore, scintigraphic results have to be matched with other radiological results and the clinical history of the patient. Putting all these information together, a diagnosis can frequently be established, by which bone scintigraphy can add important information of the metabolic level of the investigated pathologies.

> **Take-Home Message**
>
> Bone scintigraphy with technetium-99m (Tc-99m)-labeled diphosphonates is a sensitive method to image metabolic changes in bone; however, pathological tracer accumulation lacks specificity. In comparison to other imaging modalities of bone such as computed tomography and magnetic resonance imaging, bone scintigraphy is thus rather complementary to these morphologically orientated methods.

References

1. Avioli LV, Krane SM, editors. Metabolic bone disease and clinically related disorders. 3rd ed. San Diego: Academic Press; 1998.
2. Holder LE, Fogelman I, Collier BD, editors. An Atlas of planar and SPECT bone scans. 2nd ed. London: Martin Dunitz; 2000.
3. Elgazzar AH, editor. Orthopedic nuclear medicine. Berlin: Springer; 2004.
4. Hahn K, Bockisch A. Skelettsystem. In: Büll U, Schicha H, Biersack HJ, Knapp WH, Reiners C, Schober O, editors. Nuklearmedizin. Thieme, Stuttgart, 1999.
5. Mikosch P. Die Knochenszintigraphie in der Diagnostik metabolischer Knochenerkrankungen. Wien Med Wochenschr. 2004;154:119–26.
6. Ryan PJ, Fogelman I. Bone scintigraphy in metastatic bone disease. Sem Nucl Med. 1997;27:291–305.
7. Mari C, Catafau A, Carrio I. Bone scintigraphy and metabolic disorders. Q J Nucl Med. 1999;43:259–67.
8. Hendler A, Hershkop M. When to use bone scintigraphy. It can reveal things other studies cannot. Postgrad Med. 1998;104:54–66.
9. Cook GJ, Hannaford E, See M, Clarke SE, Fogelman I. The value of bone scintigraphy in the evaluation of osteoporotic patients with pack pain. Scand J Rheumatol. 2003;31:245–8.
10. Mikosch P, Pietschmann P, Kainberger F, Gallowitsch HJ, Lind P. Morbus Paget oder fibröse Dysplasie des Radius? Ein Fallbericht. Wien Med Wochenschr. 2001;151:295–9.

Pathophysiology of Bone Fragility

Katharina Kerschan-Schindl, Ursula Föger-Samwald, and Peter Pietschmann

6.1 Determinants of Bone Fragility – 84
6.1.1 Bone Mass – 84
6.1.2 Macroarchitecture – 85
6.1.3 Microarchitecture – 85
6.1.4 Material Properties – 86
6.1.5 Bone Turnover – 86
6.1.6 Bone Fragility: Concluding Remarks – 87

6.2 Pathophysiology of Osteoporosis – 87
6.2.1 Definition and Risk Factors – 87
6.2.2 Types of Osteoporosis – 88
6.2.3 Pathophysiology – 88
6.2.4 Postmenopausal Bone Loss – 90
6.2.5 «Senile» Bone Loss – 90

6.3 Pathophysiology of Non-osteoporotic Bone Diseases – 91
6.3.1 Paget's Disease of Bone – 91
6.3.2 Primary Hyperparathyroidism – 92
6.3.3 Rickets and Osteomalacia: Disorders of Bone Mineralization – 92
6.3.4 Chronic Kidney Disease-Mineral Bone Disorder – 94

References – 95

© Springer International Publishing AG 2017
P. Pietschmann (ed.), *Principles of Bone and Joint Research*, Learning Materials in Biosciences,
DOI 10.1007/978-3-319-58955-8_6

What You Will Learn in This Chapter
Fragility fractures are defined as fractures that occur as a result of a minimal trauma, such as a fall from standing height or less, or in the absence of an obvious trauma. When getting older, the probability to sustain a fragility fracture increases. This age-related increase is even more pronounced in women due to the sudden drop in sex steroid secretion during menopausal transition. In the first part of this chapter, you will learn how different bone-related properties including bone mass, bone microarchitecture and macroarchitecture, bone material properties, and bone metabolism contribute to bone fragility. The most prominent and most frequent disease associated with bone fragility is osteoporosis. However, also patients with various other diseases including Paget's disease, primary hyperparathyroidism, rickets, osteomalacia, and chronic kidney disease-mineral bone disorder (CKD-MBD) do have an increased risk to sustain fragility fractures. In the second and third part of this chapter, you will learn about the pathophysiological mechanisms leading to bone fragility in these diseases.

6.1 Determinants of Bone Fragility

One in three women and one in five men over the age of 50 years are predicted to suffer a fracture in their lifetime [1]. Fractures are among the most dramatic sequelae of the aging human skeleton leading to morbidity and an increased risk of mortality. As the proportion of elderly in the population worldwide is growing, not only the personal but also the economic burden increases exponentially. Our main goal must be to identify persons with increased risk of fragility fracture as early as possible to introduce fracture prophylaxis and if necessary specific treatment. In this chapter determinants of bone strength and the pathophysiology of selected metabolic bone diseases are discussed; for a review of malignant, infectious, or genetic diseases as a cause of bone fragility, the reader is referred elsewhere.

6.1.1 Bone Mass

Measurement of areal bone mineral density (BMD) by dual-energy absorptiometry (DXA) provides a quantitative assessment of mineralized bone mass at the axial and appendicular skeleton. It is the method of choice for diagnosis of osteoporosis and has been proven to be a strong predictor of fragility fractures. However, it has some limitations. When using the WHO classification, less than 40% of fractures occurred in osteoporotic women, 46% in osteopenic women, and 15% in women with a T-score > -1 [2]. Thus, the quantity of bone is just one among several indicators of bone health. It has been shown that age-related declines in bone strength are disproportionately steeper than decreases in BMD [3]. Only 40% of the vertebral fracture risk reduction and about 60% of the nonvertebral fracture risk reduction can be explained by the changes in total hip BMD induced by antiresorptive therapy [4]. Additionally, exercise induces only minor increases in BMD; thus, the reduction in fracture risk in physically active persons must be achieved by modifying other factors contributing to bone strength [5]. An approach in identifying persons at increased risk of fracture is to not solely rely on BMD measurements but to use a fracture risk assessment tool like FRAX® [6]; another approach is to quantify - as far as possible – other factors contributing to bone strength.

Pathophysiology of Bone Fragility

6.1.2 Macroarchitecture

Long bones are fashioned according to their demands. Bones' geometry continuously changes throughout the whole life span. During puberty growth in length ceases and the periosteal diameter increases, in girls but even more in boys. Periosteal bone apposition may be higher in males than in females [7]. After menopause the expansion of the periosteal diameter is supposed to increase and has been suggested to counteract to some extent the age-associated bone loss [8]. It has been shown quite a while ago that even without changes in bone mass, increases in diameter exponentially increase a bone's resistance to bending and torsion [9]. However, data are contradictory. According to one study, the median cortical thickness is a relevant predictor of femoral neck fragility [10], and according to other studies, femoral neck and femoral shaft diameter are not associated with fracture risk [11, 12]. Concerning the neck-shaft angle, data are also diverse [11, 12], but the longer the hip axis length, the higher the hip fracture risk [11, 12].

6.1.3 Microarchitecture

Trabecular Microarchitecture

Trabecular microarchitecture is characterized by the number of trabeculae, their average thickness, the distance between trabeculae, and the degree of connection to each other. Trabecular bone volume (TBV) is supposed to decrease by about 27% in the distal radius from the age of 20–90 years; that holds true for females as well as for males [13]. However, there are some gender-specific differences: women display a decrease in trabecular number with consequent increase in trabecular separation, whereas in men mainly trabecular thickness decreases [13]. Patients with previous fractures had higher trabecular separation and a lower interconnectivity index than those without fractures [14, 15]. Trabeculae do not only become thinner, they also shift from a platelike to a rodlike shape [5], and this conversion has been shown to be significantly associated with bone fragility [16].

Cortical Microarchitecture

Thinking of the fact that 80% of bone is cortical and that most fractures in old age are nonvertebral arising in mainly cortical areas, one realizes the importance of cortical integrity concerning bone strength. Cortical bone is not compact; it comprises 70% mineralized bone matrix and 30% extracellular fluid-filled void volume, the central canals of the Harversian and Volkmann channels [17, 18]. These canals provide a surface for osteoclasts to remove bone and, thus, increase the diameter of the channel consequently leading to cortical porosity which is known to compromise bone strength. Compared to young individuals in aged and even more in osteoporotic patients, osteonal wall thickness is decreased, and the number of osteocytes per osteon reduced leading to inferior osteonal strength [19]. By age 80 years, intracortical surfaces are supposed to exceed trabecular surfaces; cortical bone becomes more or less trabecularized, and cortical bone loss exceeds by far bone loss in the trabecular compartment [20]. Elderly people have a higher cortical porosity compared to younger subjects [21]. A study by Bala et al. showed that in women with osteoporosis, porosity is captured by areal BMD measurements, whereas in women with osteopenia cortical porosity is associated with forearm fractures

[22]. Thus, assessment of microstructure may improve sensitivity and specificity in identifying osteopenic women at risk for fracture. An autopsy study showed that older subjects did not only have thinner osteonal walls but also worse osteonal strength compared to younger subjects [19].

Microdamage

Applied load may induce diffuse microdamage or linear microcracks. Heterogeneity of collagen fiber orientation, tissue mineral density, and cement lines around osteons obstruct crack progression. Microcrack density increases exponentially with increasing age [23], and unfortunately the efficacy of damage-initiated remodeling decreases with age [24]. Thus, microdamage accumulates with increasing age, and this accumulation weakens bone strength. The structure model index (SMI) reflecting the rod- versus plate-like nature of the trabeculae is supposed to be a good predictor of microdamage [23].

6.1.4 Material Properties

Bone is a composite material with two major components: type I collagen and the inorganic calcium hydroxyapatite-like crystals. Recently renewed bone is less densely mineralized than older bone where secondary mineralization (crystal enlargement) has taken place. The degree of *matrix mineralization* plays an important role in the ability to absorb energy: it increases as mineralization increases; however, if bone is already more or less fully mineralized, the ability of cortical bone to absorb energy decreases with increasing mineral content [25]. Postmenopausal women with vertebral fractures seem to have a lower mean degree of mineralization compared with postmenopausal controls [26]. Osteomalacia is characterized by a primary mineralization defect. Fluorosis, induced by ancient osteoporosis-specific treatment with fluorides, induced a secondary mineralization defect.

Any disorder that affects *collagen* structure or its function also negatively affects bone quality. Collagen fiber orientation [27] as well as abnormalities in collagen cross-links [28, 29] has been identified as important predictors in bone strength. Osteogenesis imperfecta is a disease associated with both a defect in collagen synthesis and the degree of mineralization.

6.1.5 Bone Turnover

Cellular mechanisms are responsible for the adaptation of bone. Bone modeling and remodeling modify the external size, contours of bone, and its internal architecture. Osteocytes physiologically sense bone deformation and initiate adaptive remodeling.

Rapid remodeling is associated with an increased risk of fragility fracture because of temporarily unfilled resorption sites, presence of less densely mineralized bone, and impaired isomerization and maturation of collagen [30]. Additionally, the risk of loss of trabecular connectivity is higher in a situation of increased bone turnover. Besides osteoporosis, Paget's disease, hyperparathyroidism, hyperthyroidism, and regular intake of medications like corticosteroids are examples of pathologically high bone turnover.

However, extremely low turnover (for instance, in osteopetrosis) also has a negative impact on bone stability because it seems to be related to the presence of microcracks.

Diseases affecting bone remodeling negatively influence bone strength, but on the other hand drugs that impact bone remodeling can influence a bone's resistance to fracture in a positive way.

6.1.6 Bone Fragility: Concluding Remarks

Bone has to resist the mechanical forces applied to it by gravity and muscle contraction. The ability of bone to withstand everyday loads depends on the ability of its material and structural properties to absorb energy imposed during loading and to release it when unloaded. It depends on bone mass, bone geometry, bone architecture, and material properties including microdamage. Bone is adapting throughout life, and the age-associated loss of bone strength reflects the ongoing response to change in mechanical and hormonal environment. Postmenopausal osteoporosis is the most common derangement of bone strength.

Since the material properties of bone are influenced by many factors which are interrelated, a multicomponent strategy is needed to increase a bone's ability to withstand forces of daily life, thus reducing the risk of fragility fractures. We also have to keep in mind that skeletal trauma in the elderly is principally related to falls. In the frail elderly, both skeletal fragility and fall risk are important determinants of fracture risk because when striking the ground, the impact on bone often exceeds its load-bearing capacity.

6.2 Pathophysiology of Osteoporosis

6.2.1 Definition and Risk Factors

Osteoporosis is a multifactorial skeletal disease characterized by impaired bone mass and bone microarchitecture leading to decreased bone strength and consequently to an increased risk of fragility fractures [31]. The main clinical manifestations of osteoporosis are fractures of the vertebrae, the hip, and the wrist. Less «classical» are, i.e., fractures of the humerus, the trochanter, or the ribs [31]. According to the World Health Organization (WHO), osteoporosis is operationally defined as bone density that is 2.5 standard deviations (SD) below the mean for young healthy adults of the same sex and which is referred to as a T-score below −2.5 [32]. Individuals with a T-score between −1 and −2.5 are defined as being osteopenic and at increased risk of developing osteoporosis [32]. However, almost 50% of fractures among postmenopausal women occur in this group [2]. Thus, besides conventional risk fracture assessment based on bone density, alternative computer-based fracture risk assessment tools like FRAX®, additionally considering other risk factors than bone density were developed [6]. The main risk factors and independent predictors of osteoporotic fractures include female gender, advanced age, ethnicity, personal and parental fracture history, low body weight, immobilization or extended sedentary periods, certain medications (i.e., long-term use of corticosteroids), and lifestyle factors like nutrition, cigarette smoking, or excessive alcohol consumption [33]. Also chronic inflammatory diseases such as rheumatoid arthritis, chronic diseases associated with malabsorption like Crohn's disease, and chronic diseases that increase the propensity to fall like dementia or Parkinson's disease influence osteoporotic fracture risk [33].

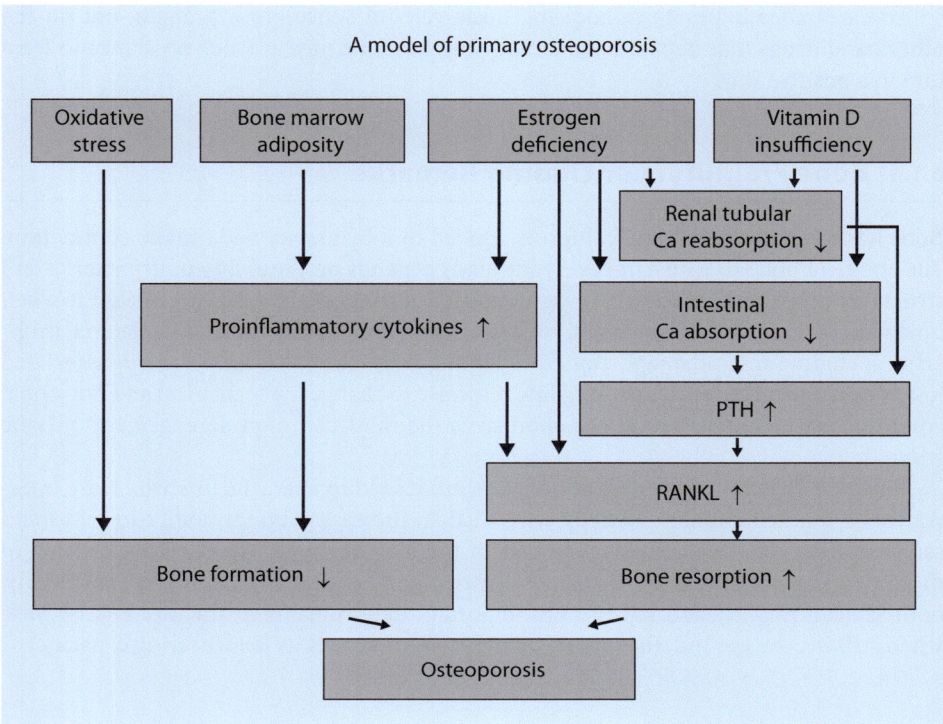

Fig. 6.1 A model of the pathogenesis of primary osteoporosis (Modified from Rauner et al. [43])

6.2.2 Types of Osteoporosis

Osteoporosis is usually classified as primary or secondary osteoporosis. Primary osteoporosis is associated with the aging process and includes rapid bone loss due to the postmenopausal decline in estrogen levels (postmenopausal bone loss) and a phase of slower bone loss due to various factors like age-associated inflammatory processes, calcium and vitamin D insufficiency, or increased parathyroid hormone (PTH) levels («senile» bone loss) (Fig. 6.1). Secondary osteoporosis is diagnosed when a secondary underlying condition (i.e., certain medical conditions or treatments) accounts for the bone loss [34]. A list of the most important causes of secondary osteoporosis is given in Table 6.1.

6.2.3 Pathophysiology

Bone fragility results first from the failure to achieve optimal peak bone mass and strength in the process of skeletal growth and second from age-related changes in bone remodeling leading to accelerated bone loss due to an imbalanced bone resorption by osteoclasts and bone formation by osteoblasts [35]. In contrast to bone modeling, which is responsible for changes in size and shape of bone and describes the formation of new bone material without preceding bone resorption, bone remodeling describes the process of concomitant resorption and replacement of bone at the same location [17]. The primary function of bone remodeling is the maintenance of skeletal integrity by replacing old and damaged

Table 6.1 Causes of secondary osteoporosis

Lifestyle	Alcoholism Smoking Immobilization
Medications	Glucocorticoids Heparin Antiepileptics
Endocrine, metabolic and systemic diseases	Diabetes mellitus Hyperthyroidism Hypogonadism Hypercalciuria Malabsorption syndrome Rheumatoid arthritis Chronic obstructive pulmonary disease (COPD) Inflammatory bowel disease Chronic liver disease
Surgery	Organ transplantation Bariatric surgery

bone matrix and the maintenance of serum calcium levels by providing calcium from the skeleton [33]. Bone remodeling takes place in spatially discrete foci, in bone multicellular units (BMUs) [36] and, in young adulthood, is a balanced process of focal resorption of bone by osteoclasts followed by replacement with the same volume of newly formed bone matrix by osteoblasts. However, as age advances, both the volume of bone resorbed by osteoclasts and the volume of bone newly formed by osteoblasts within each BMU decrease [17]. Since the volume of the newly formed bone decreases more than the volume of the resorbed bone, the net balance within each BMU becomes negative [17]. In postmenopausal women, bone loss and structural decay resulting from this negative BMU balance are potentiated by an additional abnormality. After menopause, the remodeling rate reflected by the remodeling activation frequency – the probability that a new remodeling cycle will be initiated at any point on the bone surface – increases [17]. Thus, the small imbalance within each BMU is magnified and leads to rapid bone loss and structural decay within a short time. Additionally, decreased periosteal apposition after completion of longitudinal growth and epiphyseal closure contributes to decreased biomechanical strength of long bones [33].

The fine-tuned communication between osteoclasts, osteoblasts, and surrounding cells of the bone marrow niche in bone remodeling has been established to be primarily regulated by three molecules of the tumor necrosis factor (TNF) family, the receptor activator of NF-κB (RANK) on osteoclasts, its ligand RANKL, and OPG, the decoy receptor of RANKL [37]. Circulating hormones and locally produced growth factors primarily modulating the remodeling activation frequency include estrogens, androgens, vitamin D, parathyroid hormone (PTH), parathyroid hormone-related peptide (PTHrP), IGF-I, immunoreactive growth hormone I (IGH-I), transforming growth factor β (TGF-β), interleukins (ILs), prostaglandins, and members of the tumor necrosis factor (TNF) superfamily [33].

6.2.4 Postmenopausal Bone Loss

Due to declining estrogen levels after menopause, women are at higher risk of developing osteoporosis. Menopause has been established to be associated with increasing amounts of bone lost within each BMU and with an increasing rate of bone remodeling [35]. Whereas markers of bone resorption increase by 90%, markers of bone formation increase only by 45% after menopause [38]. Cellular and molecular mechanisms underlying estrogen deficiency-dependent bone loss are increasingly well understood. Estrogen signals through two receptors, the estrogen receptor α (ERα) and the estrogen receptor β (ERβ). Both receptors are expressed on osteoclasts, osteoblasts, osteocytes, bone marrow stromal cells, and immune cell, but ERα appears to predominantly mediate estrogen's actions on bone [39]. Skeletal effects of estrogen may be subdivided into direct activities mediated by estrogen receptors on osteoblasts and osteoclasts and indirect activities mediated by estrogen receptors on other cell types including stromal cells and immune cells [40]. Estrogen deficiency, through direct and indirect activities, leads to increased osteoclast recruitment, formation, and activation and increased osteoclast life span. These events are mediated by an upregulation of RANKL and/or a downregulation OPG leading to RANKL/OPG ratios favoring bone resorption. Indirect effects of estrogen on bone very often are mediated by cytokines and inflammatory mediators released by cells of the immune system, suggesting that the postmenopausal phase is associated with a progressive proinflammatory status [41–43]. Especially T-cells are thought to play an important role in estrogen-driven bone loss [44]. Bone-resorbing cytokines and inflammatory mediators released by cells of the immune system and influenced by estrogen deficiency include IL-1, IL-6, TNF-α, macrophage colony-stimulating factor (M-CSF), and prostaglandins [45]. Estrogen also affects the longevity of osteoblasts. However, in contrast to osteoclasts, the life span of osteoblasts is decreased under the influence of estrogen [45]. Interestingly, also bone loss in aging men has been shown to be associated more closely to low estrogen levels rather than low androgen levels [45].

6.2.5 «Senile» Bone Loss

Besides gonadal sex steroid deficiency, reactive secondary hyperparathyroidism is thought to be the major driving force of age-related bone loss in men and women. Reasons for increased PTH secretion with age are vitamin D deficiency, which is common in elderly subjects, and a chronic negative calcium balance as a consequence of decreased intestinal calcium absorption and renal tubular calcium reabsorption triggered by long-standing estrogen deficiency [45]. Additionally, a progressive proinflammatory status referred to as «inflammaging» [46] may contribute to increased bone resorption by increased levels of inflammatory cytokines and mediators [47]. Decreased bone formation seen with increasing age is associated with lower numbers of osteoblasts and higher numbers of adipocytes. Osteoblasts and adipocytes share a common precursor cell, the mesenchymal stem cell. Adipogenesis therefore increases at the expense of osteoblastogenesis [48]. Systemically, decreased bone formation is attributed to decreased production of growth factors and/or decreased growth hormone (GH) and insulin-like growth factor I (IGF-I) levels [45]. Only recently, cumulating evidence suggests an important role of oxidative stress in age-related bone loss [49, 50].

6.3 Pathophysiology of Non-osteoporotic Bone Diseases

6.3.1 Paget's Disease of Bone

Paget's disease is an age-related metabolic bone disease «characterized by focal abnormalities of increased and disorganized bone turnover affecting one or more skeletal sites» [51, 52]. Although the disease can involve almost any bone, the most frequently affected sites are the pelvis, femur, spine, skull, and tibia [53]. Generally Paget's disease is considered to be the second most frequent metabolic bone disease (after osteoporosis); nevertheless, there is a significant geographic variation in the prevalence of the disease. The prevalence is highest in Western Europe (in particular in Britain), North America, and Australia; the disease is rare in Scandinavia, Asia, and Africa [53]. In autopsy series a prevalence of 3% has been reported [54]. Poor et al. assessed the prevalence of Paget's diseases in eight European towns and found a prevalence of 0.3% in subjects aged ≥55 years [55]. The authors also reported on a significant decline of the incidence; similar declines had been reported in studies from Britain and the United States (for review, see [56]).

Paget's disease of bone has a relatively strong genetic component: in 15–40% of the cases, there is a positive family history. In first-degree relatives of patients with Paget's disease, the prevalence of the disease is seven- to tenfold higher than in the background population [52, 54].

Paget's disease is a paradigm of an osteoclast-mediated disorder [56]. In the affected bones, the number and activity of osteoclasts are dramatically increased; moreover, Pagetic osteoclasts are enlarged, have an abnormally high number of nuclei, and often contain nuclear inclusion bodies. The exact mechanisms of osteoclast pathologies in Paget's disease until now have not been fully clarified. In vitro studies had suggested that Pagetic osteoclast precursors demonstrate an enhanced sensitivity to RANKL; some – but not all authors – assumed that an increased interleukin-6 expression is responsible for the osteoclast abnormalities. In patients with Paget's disease, increased serum levels of RANKL, macrophage colony-stimulating factor, or dickkopf-1 have been reported (for review, see [52]). Mutations of the Sequestosome 1 gene (SQSTM1) also appear to account for excessive bone resorption in Paget's disease: under normal conditions, p62, the gene product of SQSTM1, is involved in the suppression of RANK signaling. Mutations in certain regions of p62 result in an activation of RANK signaling and consequently in an excessive generation and activation of osteoclasts.

Since decades a viral etiology of Paget's disease has been discussed; among others, measles virus and canine distemper virus were assumed to play a pathophysiologic role. Traditionally the presence of nuclear inclusion bodies has been used as an argument for the viral etiology; nevertheless, highly sensitive contemporary molecular techniques do not support the virus hypothesis of Paget's disease [52, 57].

As a consequence of osteoclast overactivity, the earliest phase of the disease is osteolytic. In the following phase, there is a (presumably compensatory) massive increase of bone formation; the excessively high speed of bone remodeling leads to the formation of woven bone with inferior biomechanical properties. Histologically Pagetic bone is characterized by a chaotic mosaic of woven and lamellar bone [53, 54]. A comprehensive histomorphometric study in patients with Paget's disease revealed significant increases of trabecular bone volume, the number of trabeculi, the osteoid volume, and the number of osteoclasts and osteoblasts [58].

Although patients with Paget's disease may be asymptomatic, the most common symptom is bone pain. Complications of the disease include bone deformities, fractures, secondary osteoarthritis, cardiovascular complications, and neoplastic degeneration [54, 59].

6.3.2 Primary Hyperparathyroidism

Primary hyperparathyroidism is «a generalized disorder of calcium, phosphate and bone metabolism due to an increased secretion of PTH» [60]. The incidence of primary hyperparathyroidism is approximately 1 in 500 to 1 in 1000; three quarters of the patients are women. In 80–85% of the cases, the cause of the diseases is a solitary adenoma of the parathyroid gland; in 10–15% there is a hyperplasia of all (four) parathyroid glands. In rare cases primary hyperparathyroidism is due to multiple adenomas or parathyroid cancer [61].

Major actions of PTH are the stimulation of bone resorption, the retention of calcium by the kidney, and an increase of renal phosphate excretion. PTH also stimulates the 1-alpha-hydroxylase and thereby indirectly promotes intestinal calcium absorption [62]. Since in primary hyperparathyroidism the feedback regulatory mechanisms of extracellular calcium on PTH secretion are lost, excessive PTH concentrations result in hypercalcemia and hypophosphatemia. Bone turnover in primary hyperparathyroidism is characterized by an increased osteoclast activity and an increased bone formation; since (in most cases) the activity of the osteoclasts is higher than those of the osteoblasts, bone mineral density typically is decreased, particularly at cortical sites [63, 64]. Studies using high-resolution peripheral quantitative computed tomography in primary hyperparathyroidism demonstrated that microstructural abnormalities are present at the cortical and trabecular compartment and thus contribute to our understanding of the increased fracture risk [65, 66].

The clinical presentation of patients with primary hyperparathyroidism has significantly changed over time. Historically patients were highly symptomatic and presented with a specific skeletal disorder («osteitis fibrosa cystica»), nephrolithiasis, and gastrointestinal and neuromuscular manifestations (including myopathy) [61]. Today many patients are asymptomatic and are identified by routine calcium or bone density testing [67].

6.3.3 Rickets and Osteomalacia: Disorders of Bone Mineralization

Bone is composed of an organic matrix and a mineral phase; hydroxyapatite $[Ca_{10}(PO_4)_6(OH)_2]$ is the major constituent of the mineral phase. In rickets and osteomalacia, mineralization of the organic matrix («osteoid») is impaired; histologically both diseases are characterized by wide osteoid seams (◘ Fig. 6.2). The term «rickets» refers to defective mineralization of the growing skeleton, whereas «osteomalacia» is the adult equivalent of rickets. The most important causes of rickets and osteomalacia are vitamin D or phosphate deficiency (◘ Table 6.2).

Vitamin D_3 is a steroid hormone that can be synthesized endogenously in the skin as a result of ultraviolet light exposure; alternatively, vitamin D is also absorbed in the intestine from the diet. To be fully biologically active, two hydroxylation steps are necessary: 25-hydroxylation of vitamin D occurs in the liver and 1-α-hydroxylation in the kidney.

Pathophysiology of Bone Fragility

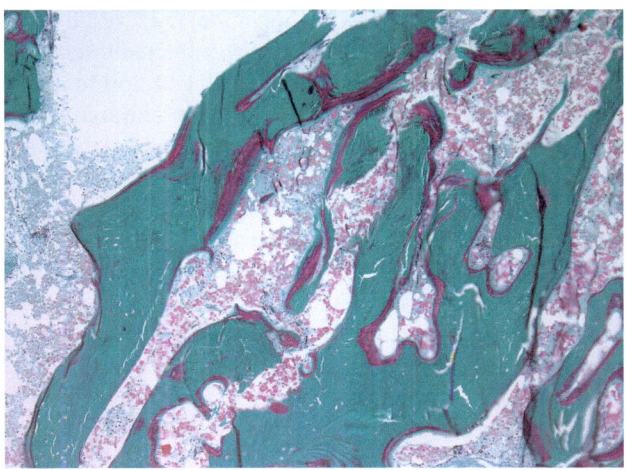

Fig. 6.2 Microscopic appearance of a bone biopsy from a patient with osteomalacia (Goldner's stain). Mineralized bone matrix is stained *green*; abnormally wide osteoid seams are stained *red* (Courtesy of Prof. Dr. Irene Sulzbacher, Department of Clinical Pathology; Medical University of Vienna)

Table 6.2 Causes of rickets and osteomalacia

Causes	Examples
Impaired vitamin D action	
Vitamin D deficiency	Reduced ultraviolet light exposure Nutritional deficiency Malabsorption
Impaired 25-hydroxylation	Liver disease
Impaired 1alpha-hydroxylation	Renal failure 1alpha-hydroxylase mutation
Target organ resistance	Vitamin D receptor mutation
Phosphate deficiency	
Decreased availability	Nutritional deficit
Reduced renal tubular phosphate reabsorption	X-linked hypophosphatemia rickets Tumor-induced osteomalacia
Acidosis	Renal tubular acidosis
Other mineralization defects	Inhibitors of mineralization (etidronate, fluoride) Hypophosphatasia

Modified from Bringhurst et al. [68] and Le Boff [70]

Major actions of 1,25(OH)$_2$vitamin D$_3$ include the promotion of calcium and phosphate absorption from the gut, the tubular reabsorption of calcium and phosphate, and the stimulation of bone resorption (via the upregulation of RANKL) [68].

Vitamin D deficiency due to nutritional deficits and/or decreased sunlight exposure is the most frequent cause of rickets and osteomalacia; nevertheless also disturbances of the activation or action of the hormone may result in impaired bone mineralization (◘ Table 6.2). In vitamin D deficiency, impaired intestinal absorption of calcium results in hypocalcemia, and in consequence – in addition to impaired bone mineralization – secondary hyperparathyroidism develops. In children with rickets due to vitamin D deficiency, an expansion of the epiphyseal growth plate, skeletal deformities (e.g., bow legs), swellings of the costochondral junction («rickety rosary»), and fractures can be observed [69]. Symptoms of osteomalacia include skeletal deformities, cortical stress fractures («Looser zones»), and proximal myopathy [70, 71].

Fibroblast growth factor-23 (FGF-23) is expressed in osteocytes, osteoblasts, and bone lining cells and (by the inhibition of tubular sodium-phosphate transporters) stimulates renal phosphate excretion. Moreover, FGF-23 inhibits the activity of the 1-α-hydroxylase and thereby the final activation of vitamin D. Due to its activity as a «phosphatonin,» FGF-23 is involved in the pathogenesis of several disorders of bone mineralization such as X-linked hypophosphatemic rickets and tumor-induced osteomalacia [72, 73].

X-linked hypophosphatemic rickets (XLH) is caused by a mutation of the PHEX (phosphate-regulating gene with homologies to endopeptidases on the X-chromosome) gene. It appears that the defect of the PHEX gene inappropriately increases FGF-23 expression; in consequence urinary phosphate excretion is enhanced and results in hypophosphatemia. Clinical manifestations of XLH include growth retardation, limb deformities, the rachitic rosary, and dental defects [72, 73].

Tumor-induced osteomalacia (TIO) is a paraneoplastic syndrome associated with mesenchymal tumors. Biochemically TIO is characterized by phosphaturia and hypophosphatemia but normal calcium and PTH levels. TIO is assumed to be caused by an excessive production of FGF-23 (or other phosphatonins) by mesenchymal tumors.

6.3.4 Chronic Kidney Disease-Mineral Bone Disorder

The term chronic kidney disease-mineral bone disorder (CKD-MBD) refers to a complex syndrome frequently observed in chronic renal failure [74]. The pathogenesis of CKD-MBD is multifactorial and includes hyperphosphatemia, 1,25(OH)$_2$vitamin D deficiency, hypocalcemia, hyperparathyroidism, and FGF-23. Histologically the spectrum of renal osteodystrophy (the skeletal component of CKD-MBD) is very broad; patients can present with high bone turnover (secondary hyperparathyroidism), low bone turnover (adynamic bone disease), osteomalacia, or mixed uremic osteodystrophy. Clinical consequences of renal osteodystrophy are localized or diffuse bone pain, skeletal deformities, and fractures [75].

> **Take-Home Message**
> - Besides bone mass, bone fragility is determined by various other bone-related properties including bone microarchitecture, bone macroarchitecture, bone material properties, and bone metabolism.
> - Osteoporosis is the most frequent disease associated with bone fragility.
> - Various other diseases, e.g., Paget's disease, rickets, osteomalacia, primary hyperparathyroidism, or chronic kidney disease-mineral bone disorder (CKD-MBD), are associated with bone fragility.

References

1. Johnell O, Kanis J. Epidemiology of osteoporotic fractures. Osteoporos Int. 2005;16(Suppl 2):TaS3–7.
2. Boutroy S, Hans D, Sornay-Rendu E, Vilayphiou N, Winzenrieth R, Chapurlat R. Trabecular bone score improves fracture risk prediction in non-osteoporotic women: the OFELY study. Osteoporos Int. 2013;24(1):77–85.
3. Keaveny TM, Kopperdahl DL, Melton LJ 3rd, Hoffmann PF, Amin S, Riggs BL, et al. Age-dependence of femoral strength in white women and men. J Bone Miner Res. 2010;25(5):994–1001.
4. Jacques RM, Boonen S, Cosman F, Reid IR, Bauer DC, Black DM, Eastell R. Relationship of changes in total hip bone mineral density to vertebral and nonvertebral fracture risk in women with postmenopausal osteoporosis treated with once-yearly zoledronic acid 5 mg: the HORIZON-pivotal fracture trial (PFT). J Bone Miner Res. 2012;27(8):1627–34.
5. Fonseca H, Moreira-Goncalves D, Coriolano HJA, et al. Bone quality: the determinants of bone strength and fragility. Sports Med. 2013;44(1):37–53.
6. Kanis JA, Johnell O, Oden A, Johansson H, McCloskey E. FRAX and the assessment of fracture probability in men and women from the UK. Osteoporos Int. 2008;19(4):385–97.
7. Seeman E. Pathogenesis of bone fragility in women and men. Lancet. 2002;359(9320):1841–50.
8. Ahlborg HG, Johnell O, Turner CH, Rannevik G, Karlsson MK. Bone loss and bone size after menopause. N Engl J Med. 2003;349(4):327–34.
9. Turner CH, Burr DB. Basic biomechanical measurements of bone: a tutorial. Bone. 1993;14(4):595–608.
10. Kersh ME, Pandy MG, Bui QM, Jones AC, Arns CH, Knackstedt MA, et al. The heterogeneity in femoral neck structure and strength. J Bone Miner Res. 2013;28(5):1022–8.
11. Faulkner KG, Cummings SR, Black D, Palermo L, Gluer CC, Genant HK. Simple measurement of femoral geometry predicts hip fracture: the study of osteoporotic fractures. J Bone Miner Res. 1993;8(10):1211–7.
12. Gnudi S, Sitta E, Pignotti E. Prediction of incident hip fracture by femoral neck bone mineral density and neck-shaft angle: a 5-year longitudinal study in post-menopausal females. Br J Radiol. 2012;85(1016):e467–73.
13. Khosla S, Riggs BL, Atkinson EJ, Oberg AL, McDaniel LJ, Holets M, et al. Effects of sex and age on bone microstructure at the ultradistal radius: a population-based noninvasive in vivo assessment. J Bone Miner Res. 2006;21(1):124–31.
14. Legrand E, Chappard D, Pascaretti C, Duquenne M, Krebs S, Rohmer V, et al. Trabecular bone microarchitecture, bone mineral density, and vertebral fractures in male osteoporosis. J Bone Miner Res. 2000;15(1):13–9.
15. Milovanovic P, Djonic D, Marshall RP, Hahn M, Nikolic S, Zivkovic V, et al. Micro-structural basis for particular vulnerability of the superolateral neck trabecular bone in the postmenopausal women with hip fractures. Bone. 2012;50(1):63–8.
16. Pialat JB, Vilayphiou N, Boutroy S, Gouttenoire PJ, Sornay-Rendu E, Chapurlat R, et al. Local topological analysis at the distal radius by HR-pQCT: application to in vivo bone microarchitecture and fracture assessment in the OFELY study. Bone. 2012;51(3):362–8.
17. Seeman E. Age- and menopause-related bone loss compromise cortical and trabecular microstructure. J Gerontol A Biol Sci Med Sci. 2013;68(10):1218–25.
18. Cooper DM, Thomas CD, Clement JG, Turinsky AL, Sensen CW, Hallgrimsson B. Age-dependent change in the 3D structure of cortical porosity at the human femoral midshaft. Bone. 2007;40(4):957–65.
19. Bernhard A, Milovanovic P, Zimmermann EA, Hahn M, Djonic D, Krause M, et al. Micro-morphological properties of osteons reveal changes in cortical bone stability during aging, osteoporosis, and bisphosphonate treatment in women. Osteoporos Int. 2013;24(10):2671–80.
20. Zebaze RM, Ghasem-Zadeh A, Bohte A, Iuliano-Burns S, Mirams M, Price RI, et al. Intracortical remodelling and porosity in the distal radius and post-mortem femurs of women: a cross-sectional study. Lancet. 2010;375(9727):1729–36.
21. Nicks KM, Amin S, Atkinson EJ, Riggs BL, Melton LJ 3rd, Khosla S. Relationship of age to bone microstructure independent of areal bone mineral density. J Bone Miner Res. 2012;27(3):637–44.
22. Bala Y, Zebaze R, Ghasem-Zadeh A, Atkinson EJ, Iuliano S, Peterson JM, et al. Cortical porosity identifies women with osteopenia at increased risk for forearm fractures. J Bone Miner Res. 2014;29(6):1356–62.
23. Arlot ME, Burt-Pichat B, Roux JP, Vashishth D, Bouxsein ML, Delmas PD. Microarchitecture influences microdamage accumulation in human vertebral trabecular bone. J Bone Miner Res. 2008;23(10):1613–8.

24. Waldorff EI, Goldstein SA, McCreadie BR. Age-dependent microdamage removal following mechanically induced microdamage in trabecular bone in vivo. Bone. 2007;40(2):425–32.
25. Graeff C, Timm W, Nickelsen TN, Farrerons J, Marin F, Barker C, et al. Monitoring teriparatide-associated changes in vertebral microstructure by high-resolution CT in vivo: results from the EUROFORS study. J Bone Miner Res. 2007;22(9):1426–33.
26. Boivin G, Vedi S, Purdie DW, Compston JE, Meunier PJ. Influence of estrogen therapy at conventional and high doses on the degree of mineralization of iliac bone tissue: a quantitative microradiographic analysis in postmenopausal women. Bone. 2005;36(3):562–7.
27. Martin RB, Ishida J. The relative effects of collagen fiber orientation, porosity, density, and mineralization on bone strength. J Biomech. 1989;22(5):419–26.
28. Oxlund H, Barckman M, Ortoft G, Andreassen TT. Reduced concentrations of collagen cross-links are associated with reduced strength of bone. Bone. 1995;17(4 Suppl):365S–71S.
29. Viguet-Carrin S, Follet H, Gineyts E, Roux JP, Munoz F, Chapurlat R, et al. Association between collagen cross-links and trabecular microarchitecture properties of human vertebral bone. Bone. 2010;46(2):342–7.
30. Seeman E, Delmas PD. Bone quality--the material and structural basis of bone strength and fragility. N Engl J Med. 2006;354(21):2250–61.
31. Rachner TD, Khosla S, Hofbauer LC. Osteoporosis: now and the future. Lancet. 2011;377(9773):1276–87.
32. Report of a WHO Study Group. Assessment of fracture risk and its application to screening for postmenopausal osteoporosis. World Health Organ Tech Rep Ser. 1994;843:1–129.
33. Lindsay R, Cosman F. Osteoporosis. In: Longo DL, Kasper DL, Jameson JL, Fauci AS, Hauser SL, Loscalzo J, editors. Harrison's principles of internal medicine. New York: Mc Graw Hill; 2012.
34. Painter SE, Kleerekoper M, Camacho PM. Secondary osteoporosis: a review of the recent evidence. Endocr Pract. 2006;12(4):436–45.
35. Raisz LG. Pathogenesis of osteoporosis: concepts, conflicts, and prospects. J Clin Invest. 2005;115(12):3318–25.
36. Hattner R, Epker BN, Frost HM. Suggested sequential mode of control of changes in cell behaviour in adult bone remodelling. Nature. 1965;206(983):489–90.
37. Leibbrandt A, Penninger JM. RANK/RANKL: regulators of immune responses and bone physiology. Ann N Y Acad Sci. 2008;1143:123–50.
38. Garnero P, Sornay-Rendu E, Chapuy MC, Delmas PD. Increased bone turnover in late postmenopausal women is a major determinant of osteoporosis. J Bone Miner Res. 1996;11(3):337–49.
39. Riggs BL, Khosla S, Melton LJ 3rd. Sex steroids and the construction and conservation of the adult skeleton. Endocr Rev. 2002;23(3):279–302.
40. Sipos W, Pietschmann P, Rauner M, Kerschan-Schindl K, Patsch J. Pathophysiology of osteoporosis. Wien Med Wochenschr. 2009;159(9–10):230–4.
41. Pacifici R. Estrogen deficiency, T cells and bone loss. Cell Immunol. 2008;252(1–2):68–80.
42. Pietschmann P, Rauner M, Sipos W, Kerschan-Schindl K. Osteoporosis: an age-related and gender-specific disease--a mini-review. Gerontology. 2009;55(1):3–12.
43. Rauner M, Sipos W, Thiele S, Pietschmann P. Advances in osteoimmunology: pathophysiologic concepts and treatment opportunities. Int Arch Allergy Immunol. 2013;160(2):114–25.
44. Pacifici R. Mechanism of estrogen action in bone. In: Bilezikian JP, Raisz LG, Martin JT, editors. Principles of bone biology. Third ed. New York: Academic Press; 2008. p. 921–33.
45. Clarke BL, Khosla S. Physiology of bone loss. Radiol Clin N Am. 2010;48(3):483–95.
46. Franceschi C, Bonafe M, Valensin S, Olivieri F, De Luca M, Ottaviani E, et al. Inflamm-aging. An evolutionary perspective on immunosenescence. Ann N Y Acad Sci. 2000;908:244–54.
47. Pietschmann P, Mechtcheriakova D, Meshcheryakova A, Föger-Samwald U, Ellinger I. Immunology of osteoporosis: a mini-review. Gerontology. 2016;62(2):128–37.
48. Duque G, Troen BR. Understanding the mechanisms of senile osteoporosis: new facts for a major geriatric syndrome. J Am Geriatr Soc. 2008;56(5):935–41.
49. Manolagas SC. From estrogen-centric to aging and oxidative stress: a revised perspective of the pathogenesis of osteoporosis. Endocr Rev. 2010;31(3):266–300.
50. Föger-Samwald U, Vekszler G, Hörz-Schuch E, Salem S, Wipperich M, Ritschl P, Mousavi M, Pietschmann P. Molecular mechanisms of osteoporotic hip fractures in elderly women. Exp Gerontol. 2016;73:49–58.
51. Kanis JA. Pathophysiology and treatment of Paget's disease of bone. London: Martin Dunitz; 1992.

52. Ralston SH, Layfield R. Pathogenesis of Paget disease of bone. Calcif Tissue Int. 2012;91(2):97–113.
53. Siris ES, Roodman GD. Paget's disease of bone. In: Rosen CJ, Bouillon R, Compston JE, Rosen V, editors. Primer on the metabolic bone diseases and disorders of mineral metabolism. Oxford: Wiley-Blackwell; 2013. p. 659–68.
54. Favus MJ, Vokes TJ. Paget's disease and other dysplasias of bone. In: Longo DL, Kasper DL, Jameson JL, Fauci AS, Hauser SL, Loscalzo J, editors. Harrison's principles of internal medicine. New York: Mc Graw Hill; 2012. p. 3136–44.
55. Poor G, Donath J, Fornet B, Cooper C. Epidemiology of Paget's disease in Europe: the prevalence is decreasing. J Bone Miner Res. 2006;21(10):1545–9
56. Nebot Valenzuela E, Pietschmann P. Epidemiology and pathology of Paget's disease of bone – a review. Wien Med Wochenschr. 2017;167:2–8.
57. Helfrich MH, Hobson RP, Grabowski PS, Zurbriggen A, Cosby SL, Dickson GR, et al. A negative search for a paramyxoviral etiology of Paget's disease of bone: molecular, immunological, and ultrastructural studies in UK patients. J Bone Miner Res. 2000;15(12):2315–29.
58. Seitz S, Priemel M, Zustin J, Beil FT, Semler J, Minne H, et al. Paget's disease of bone: histologic analysis of 754 patients. J Bone Miner Res. 2009;24(1):62–9.
59. Altman RD. Paget's disease of bone. In: Coe FL, Favus MJ, editors. Disorders of bone and mineral metabolism. New York: Raven press; 1992. p. 1027–64.
60. Potts JT, Jüpper H. Disorders of the parathyroid g and and calcium homeostasis. In: Longo DL, Kasper DL, Jameson JL, Fauci AS, Hauser SL, Loscalzo J, editors. Harrison's principles of internal medicine. New York: Mc Graw Hill; 2012. p. 3096–120.
61. Silverberg SJ. Primary hyperparathyroidism. In: Rosen CJ, Bouillon R, Compston JE, Rosen V, editors. Primer on the metabolic bone diseases and disorders of mineral metabolism. Oxford: Wiley-Blackwell; 2013. p. 543–52.
62. Nissenson RA, Jüpper H. Parathyroid hormone. In: Rosen CJ, Bouillon R, Compston J, Rosen V, editors. Primer on the metabolic bone diseases and disorders of mineral metabolism. Oxford: Wiley-Blackwell; 2013. p. 208–14.
63. Kerschan-Schindl K, Riss P, Krestan C, Rauner M, Bieglmayer C, Gleiss A, et al. Bone metabolism in patients with primary hyperparathyroidism before and after surgery. Horm Metab Res. 2012;44(6):476–81.
64. Lewiecki EM, Miller PD. Skeletal effects of primary hyperparathyroidism: bone mineral density and fracture risk. J Clin Densitom. 2013;16(1):28–32.
65. Stein EM, Silva BC, Boutroy S, Zhou B, Wang J, Ucesky J, et al. Primary hyperparathyroidism is associated with abnormal cortical and trabecular microstructure and reduced bone stiffness in postmenopausal women. J Bone Miner Res. 2013;28(5):1029–40.
66. De Geronimo S, Romagnoli E, Diacinti D, D'Erasrro E, Minisola S. The risk of fractures in postmenopausal women with primary hyperparathyroidism. Eur J Endocrinol. 2006;155(3):415–20.
67. Bandeira L, Bilezikian J. Primary hyperparathyroidism. J Oral Pathol Med. 2015;44(4):239–43.
68. Bringhurst FR, Demay MB, Krane SM, Kronenberg HM. Bone and mineral metabolism in health and disease. In: Longo DL, Kasper DL, Jameson JL, Fauci AS, Hauser SL, Loscalzo J, editors. Harrison's principles of internal medicine. New York: Mc Graw Hill; 2012.
69. Wharton B, Bishop N. Rickets. Lancet. 2003;362(9393):1389–400.
70. LeBoff MS. Metabolic bone disease. In: Kelley WN, Harris ED, Ruddy S, Sledge CB, editors. Textbook of rheumatology. Philadelphia: W.B. Saunders Company; 1997. 1563–80.
71. Reddy Munagala VV, Tomar V. Images in clinical medicine. Osteomalacia N Engl J Med. 2014;370(6):e10.
72. de Menezes FH, de Castro LC, Damiani D. Hypophosphatemic rickets and osteomalacia. Arq Bras Endocrinol Metabol. 2006;50(4):802–13.
73. Ruppe MD, Jan de Beur SM. Disorders of phosphate homeostasis. In: Rosen CJ, Bouillon R, Compston J, Rosen V, editors. Primer on the metabolic bone diseases and disorders of mineral metabolism. Oxford: Wiley-Blackwell; 2013. p. 632–9.
74. Raubenheimer EJ, Noffke CE, Hendrik HD. Chronic kidney disease-mineral bone disorder: an update on the pathology and cranial manifestations. J Oral Pathol Med. 2015;44(4):239–43.
75. Hruska K, Seifert M. Pathophysiology of chronic kidney disease mineral bone disorder (CKD-MBD). In: Rosen CJ, Bouillon R, Compston J, Rosen V, editors. Primer on the metabolic bone diseases and disorders of mineral metabolism. Oxford: Wiley-Blackwell; 2013.

Fracture Healing

**From Bench to Bedside
(Physiological and Pathophysiological Aspects)**

Nicolas Haffner, Daniel Smolen, and Rainer Mittermayr

7.1 Introduction – 100

7.2 Etiology – 101

7.3 Epidemiology – 102

7.4 Principles of Fracture Treatment/Osteosynthesis – 102

7.5 Experimental Models in Fracture Repair – 104

7.6 Physiology of Fracture Healing – 108
7.6.1 Primary/Direct (Intramembranous) Healing – 108
7.6.2 Secondary/Indirect (Endosteal/Intramembranous) Healing – 109

7.7 Pathophysiology of Fracture Healing – 116
7.7.1 Classifications of Nonunions (NU) – 117
7.7.2 Molecular Alterations – 118
7.7.3 Treatment Modalities – 120

References – 123

What You Will Learn in This Chapter
This chapter will focus on principles of fracture healing. The reader is introduced to epidemiology, physiology, and pathophysiology of fracture healing, experimental models, as well as principles of conservative and operative fracture treatments.

At the end of this chapter, the reader should have a thorough understanding of theoretical basics and clinical aspects of fracture biology.

7.1 Introduction

Fracture healing is a complex physiological process, which consists of multiple steps including angiogenesis and callus formation in order to allow osseous consolidation. The inherent time course of these steps is meticulously preprogrammed in transcription and translation of different proteins. Unlike other tissues, bone has the ability to regenerate rather than to repair itself. This underlines the incredible dynamic and vitality of bone. Since the inauguration of the AO (Arbeitsgemeinschaft für Osteosynthesefragen) in 1958 in Chur, Switzerland, bone healing has been extensively investigated. Whereas in the beginning different osteosynthesis methods such as plating versus intramedullary nailing were the main focus of research, nowadays bone biology has become of increasing interest. As a consequence, osteosynthesis has focused more on bone biology as seen in developments of, e.g., minimally invasive plate osteosynthesis (MIPO). MIPO creates fracture stability without interfering with bone biology at the fracture site. Not seeing significant differences in fracture healing using different osteosynthesis materials such as titanium or steel, genetics – in particular proteins involved in fracture healing – have become of primary interest.

In order to investigate histological, biochemical, and biomechanical aspects of fracture healing, research has covered many experimental settings in terms of different animal models as well as the collection and analysis of clinical data over the years.

The word fracture derives from the Latin word *frangere*, meaning «to break,» and is defined as a disruption of cortical continuity of bone. The term cortical continuity remains critical, because incomplete fractures exist as so-called greenstick and torus fractures in children and as trabecular fractures (bone bruise or posttraumatic bone marrow edema) and fissures (disruption of only one cortex) in adults.

As Maurice Müller already postulated, a classification needs to be simple, applicable, and reliable, and ideally a therapeutic algorithm can be deduced from it. A clear descriptive classification system is important as it not only facilitates communication among clinicians but also leads to the possibility of comparing outcomes.

Fracture classification according to:		
Mechanism		
Traumatic	Fractures due to an adequate trauma	
Pathological	Fracture due to an inadequate trauma	
	Fragility/osteoporotic fractures	Fractures due to osteolysis, e.g., metastasis

Fracture Healing

Fracture classification according to:		
Insufficiency	Overuse/repetitive strain (Arndt-Schulz law)	
Periprosthetic		
Soft tissue involvement		
Open	Classification according to Gustilo and Anderson, Tscherne, AO	
	Clean	Contaminated
Closed	No skin disruption – AO	
Displacement		
Non-displaced	Displaced	
	Translated	
	Angulated	
	Rotated	
	Shortened	
	Fragmented	
	Comminuted/multi-fragmented	
Fracture pattern		
Linear		
Transverse		
Oblique		
Spiral		
Compression (wedge)		
Impaction		
Avulsion		

7.2 Etiology

Fractures occur when forces applied to the bone exceed its distinct strength. Fracture origin is influenced by extrinsic and intrinsic factors. Extrinsic factors include the mechanical load rate as well as the extent, direction, and grandeur of forces to which the bone is exposed. Intrinsic factors include the mechanical properties of the involved bone, such as the energy-absorbing capacity, Young's modulus, and strength, the latter being often clinically expressed as bone mineral density as a surrogate parameter.

Fractures arise from direct or indirect trauma. Direct trauma implies a direct force applied to the bone, whereas indirect trauma is a result of tension and compressive or rotational lever forces distant to the fracture site.

7.3 Epidemiology

According to reports of different health-care systems, we can only estimate incidences of fractures. Since not all fractures are recorded, we have to assume that numbers in general underestimate the prevalence of fractures. Approximately 5.6 million fractures occur annually in the United States (2% incidence). A peer-reviewed publication from the United States suggests that the lifetime risk of fracture is 50% for men and 33% for women [1].

Extrapolating the data of Donaldson et al. and Sahlin and Court-Brown et al. projected a worldwide fracture incidence of 9–22.8/1000/year. Gender distribution of the fracture incidence in the Scottish case series was almost equal (1.13% in men and 1.16% in women). Fractures in males, however, showed a bimodal distribution with a high incidence in young men and a second rise starting at the age of 60, whereas in women a constant rise after menopause can be observed. The increase of postmenopausal women can be interpreted as boost in fragility fractures [2].

Due to the fact that fractures account for the vast majority of trauma in developing nations, there is a constant seek for novel therapies to improve patient outcome and reduce overall costs.

7.4 Principles of Fracture Treatment/Osteosynthesis

- **Conservative Treatment**

Sir John Charnley once stated that «contrary to popular ideas, the operative treatment of fracture is much simpler than the non-operative.» Independent of the method applied for definite fracture treatment, fractures which are displaced need to be reduced prior to fixation. In conservative treatment this reduction needs to be performed in a closed fashion. After initial reduction the position has to be retained. This can be achieved either by casts/splints or traction. These fundamental principles, which are not limited to conservative treatment, can be retrieved in Lorenz Böhler's axiom of fracture care: reduction, retention, and exercise.

- **Cast/Splint**

Conservative treatment with casts or splints is still a very effective treatment especially in minimal or non-displaced fractures at the meta- or diaphyseal level. Casts can be made from thermoplastic materials such as fiberglass or out of plaster of Paris, the latter having the advantage of being more malleable and therefore being used in cases where a closed reduction is sought to be maintained. Fiberglass on the other hand is splash water-resistant, is lighter, and thus has a higher wear comfort for the patient compared to conventional plaster of Paris.

- **Traction**

According to Wolff's law of the transformation of bones (1892), traction has been widely used in the past in order to treat fractures. There are two main principles of applying traction, one of which is applying traction to the soft tissues (skin traction) and the other being traction applied to the bone (skeletal traction). Due to the constant progress in operative techniques, traction is rarely used for definite fracture treatment. In severely injured patients, however, it can be applied in order to bridge the time until definite operative treatment can be applied.

Fracture Healing

■ Surgical Therapy

Irrespective of the chosen surgical fixation method, there are two main principles of load transmission in fractured bones:
- Force transmission through the fracture/force induction: the force is transmitted from one fracture fragment to the next. This direct transmission implies that the osteosynthesis yields a certain degree of stability and that fracture fragments are in immediate vicinity and close contact.
- Force deflection around the fracture: the osteosynthesis acts as a bypass in order to allow secondary bone healing at the fracture site. Forces are transmitted over the osteosynthetic material from one fracture end to the other, avoiding any force induction at the fracture site.

The Four AO Principles
- Anatomic reduction of the fracture fragments: for the diaphysis this implies to ensure that length, angulation, and rotation are corrected; for intra-articular fractures anatomic reduction of all fragments as well as absolute stability is demanded.
- Stable fixation – absolute or relative stability depending on the fracture site: Absolute stability in cases of intra-articular fractures and relative stability at the meta- and diaphyseal level are required.
- Preservation of blood supply to the injured area of the extremity and respect for the surrounding soft tissues.
- Early range of motion (ROM) and rehabilitation.

■ Open Reduction and Internal Fixation (ORIF)

In contrast to the beginnings of fracture fixation, where anatomic reduction and absolute stability were favored, nowadays the main objective of fracture care is to minimize soft tissue damage and to avoid periosteal stripping at the fracture site. After realignment of the fracture ends, the fracture must be stabilized in order to maintain the reduction. There are multiple auxiliary devices for internal fixation.

■ Kirschner Wires (K-Wires)

K-wires are commonly used for temporary or definitive treatment of fractures. They allow maintaining reduction, resisting alignment changes in the frontal, sagittal, and transversal plane. However, they do not oppose rotation and their resistance to torque and bending forces is poor (depending on its diameter). Due to their smaller diameter, they are often employed around smaller joints (meta- and epiphyseal region) and in particular to minimize potential growth disturbances in children and adolescents where growth plates are still open.

Another classic application of K-wires combined with steel bands (cerclages) is their use in tension band osteosynthesis, for instance, of the patella or the olecranon.

■ Plates and Screws

Plates and screws are frequently applied in cases where absolute stability is required. This is the case in but not limited to the management of articular fractures. After anatomic reduction of the fracture, fragments as well as fracture ends are kept in place with screws and plates. Due to their superior stability compared to K-wires, additional casting is not mandatory, and early postoperative movement is allowed.

Depending on the anatomic region and size of the bone, the design of the plates can vary extensively. Over the years a lot of progress has been made in the development of plates to address different problems in today's facture care. The increasing age of our population leads to higher rates of osteoporosis, impeding construct stability. Therefore locking plates or fixed angled devices have been introduced to the market.

According to the AO there are five main plate functions:
- Buttress (antiglide) plates
- Compression plates
- Neutralization plates
- Tension band plate
- Bridge plates

Intramedullary Nails

At the latest from the broader propagation of intramedullary nails through Küntscher, their clinical use has been widely accepted. Intramedullary devices operate as internal splints, allowing a directed dynamic compression under weight-bearing conditions. Initially, loads are shared with the bone allowing the fracture to heal. The following different nails are available:
- Flexible or rigid
- Locked or unlocked
- Reamed or unreamed

External Fixation

In 1907, the Belgian physician Albin Lambotte introduced the technique of external fixation for fracture management. External fixation provides fast initial fracture stabilization. Depending on the respective use of Schanz screws and tubes, both main biomechanical principles of fracture care, being force deflection or force transmission at the fracture site, can be realized using an external fixator. The classic indication remains the polytraumatized patient, who has to be temporarily stabilized before definite fracture treatment can be undertaken in the so-called window of opportunity (5–12 days after the initial trauma). Due to the fact that the fracture site itself is left untouched, external fixators are also commonly applied in multi-fragmented fractures with severely impaired surrounding soft tissues (open comminuted fractures). The technique provides stability for the extremity maintaining bone length, alignment, and rotation without requiring casting. Therefore, it allows for continuous inspection as well as subsequent wound care of soft tissue structures.

7.5 Experimental Models in Fracture Repair

Over the past an increasing knowledge in understanding the process of fracture healing has accumulated, which fundamentally supports clinicians and researchers in their respective work to further elucidate subprocesses of physiological and pathophysiological bone repair as well as developing strategies to treat fractures.

To adequately address these facts, experimental models are essential. However, several key components have to be considered to apply the most suitable model to answer a specific research question.

The experimental model should best possibly reflect the following clinical scenarios:
- Normal bone healing
  - Direct fracture healing
  - Indirect fracture healing
- Abnormal bone healing
  - Fractures afflicted with high incidence of disturbed healing
  - Delayed fracture union
  - Established nonunion
    - Hypertrophic
    - Oligotrophic
    - Atrophic
    - Infected
  - Segmental bony defects
- Osseointegration

General aspects have to be taken into account prior to performing an experimental study on bone repair. Dependent on the clinical scenarios mentioned above, it has to be deliberated whether a small- or a large-scale animal model should be used. Additionally the researcher has to decide which species will be the most appropriate to answer the fundamental study question. Besides, the size and species of the animal model – with their respective advantages and limitations – age, gender, type of fracture, and fixation technique and time schedule of follow-up have to be considered in advance in order to minimize bias.

As shown by Pearce et al. and others, there are various diversified animal models in trauma/orthopedic research [3, 4]. Rodents are the most commonly utilized animals in fracture research (rats 36–38%, mice 15–26%), followed by rabbits (13–19%), dogs (9%), and primates (3%) [5]. Sheep account for 2–11% of the species used in studies of fracture healing [6].

Substantial differences among one and in between species can be found not only in the microarchitecture and composition of bone but also in the biochemistry, biomechanics, and their respective healing responses [4, 7, 8].

Due to the fact that experimental models are primarily used to further elucidate physiological and pathophysiological processes in human fracture healing, careful attention has to be drawn to the differences in anatomy, biology, and fracture healing associated with biochemical pathways between species and models, in order to optimally extrapolate the experimental results and to translate them into the clinical environment.

■ Mice

Bone maturity, defined as physeal closure in mice, is reached at an age of 5 months, measured at the animal's tibia. Their overall life expectancy ranges from 18 to 36 months, making cross comparisons to humans difficult. Usually bony union can be seen after 3–4 weeks in a simple fracture model in mice, which compares relatively long with respect to their life expectancy.

An important characteristic of small rodent models (mice and rats) is the lack of Haversian systems. Thus the bone is remodeled at the fracture site by resorption cavities. Consequently these lacunae are filled by osteoblasts as the bone heals. Therefore this model can clearly not be applied in research focusing on properties of the Haversian system.

Nevertheless, reasonable pricing, availability, and handling designate mice to be an attractive and frequently used experimental animal model. Furthermore, the growing insight in the genetic background with the subsequent opportunity of genetic knockout mutants can be used to study distinct molecular mechanisms of bone healing.

- **Rats**

Rats are frequently used to study fracture healing. Rat models usually involve long bones or the calvaria. As in mice they are cheap in purchase and housing and are quite easy to handle.

Physeal plate closure is considered at the age of 11 months with a life expectancy of 30–48 months. Under normal circumstances fracture healing in rats takes 4–5 weeks. This is mainly achieved by the external periosteal callus, with only minor medullary callus involvement.

Generally, endochondral bone formation predominates in mice as well as in rats. Rats, especially older ones that underwent ovariectomy, have reduced bone mineral density and show delayed healing of femoral fractures. Therefore they are frequently used as a model for osteoporotic fractures.

- **Rabbits**

In rabbits (New Zealand White) cessation of bone growth (physeal closure) is anticipated within 5.3–6.8 months in the tibia and femur, respectively. Following fracture expected healing is achieved at 6–7 weeks. As measured by life expectancy of 84–96 months, the repair process engages disproportionally longer time periods in comparison to humans. Remodeling in rabbits occurs quickly showing a different microstructure than seen in humans, although a Haversian system in this higher phylogenetic species is already present.

Due to its size the rabbit is used for various study designs, comprising experiments of fixation methods, such as intramedullary nailing/splinting, external fixations, or plates and screws, but also established in biophysical experiments [9]. Animal housing and costs still allow for moderate-sized groups when different protocols are used.

- **Dogs**

The dogs span of life ranges between 108 and 168 months with physeal closure at approximately 7.5 months (Greyhounds). After a simple fracture, time to fracture union can be expected within 10–13 months. Even though canine bones have similarities to human bones regarding their architecture, composition, and remodeling processes, they exhibit a combination of plexiform and lamellar bone, which cannot be seen in humans. Remodeling therefore is highly variable, and biomechanical properties differ from those of human bones.

It is to note that dogs are more susceptible to nonsteroidal anti-inflammatory drugs due to their higher gastrointestinal resorption rate. The increased oral bioavailability and the longer half-life of these drugs make ulcer formation associated with significant gastrointestinal bleeding more likely to occur. This has to be considered carefully in planning the postoperative analgesic regimen.

However, the main concern using dogs in experimental settings remains an ethical one as the dog being «men's best friend.» Additionally they are expensive and demanding to keep.

Fracture Healing

■ **Sheep**

Extensive research is accomplished in sheep to study various fixation devices in different setups. The life expectancy of sheep is around 180 months with a physeal growth plate closure at the age of approximately 17 months. The bone (cancellous, cortical, and plexiform) will usually heal after 10–14 weeks following fracture. The advantage of experiments in sheep is the similar body weight to humans along with a suitable dimension of long bones for implants. Unlike other species, sheep only rarely protect their fractured hind limbs, thus making this model suitable for fixation techniques where early weight-bearing is favored. Limitations are housing, handling, and availability issues as well as the age-dependent bone remodeling.

Alternatively to in vivo experimental models, mathematical models can add feasible means to further elucidate normal and pathological fracture repair along with the respective treatment (CIT 25 Pivonka). Moreover, we have a sincere ethical responsibility to reduce experimental in vivo experiments, by which mathematical modeling may have the potential to support research in this effort.

■ **Experimental Fracture Models**

Osteotomy – Although highly reproducible this approach has to be used with caution when traumatic fractures and their respective therapies are the main research focus. Significant differences were found in a rabbit tibial repair model using either a closed fracture model or an osteotomy [10].

To create osteotomies an open approach is inherent thus lacking fracture hematoma which is a major component in physiological fracture healing. Another crucial point to be considered in creating osteotomies is the developing heat caused by the oscillating saw which undoubtedly influences local cells and subsequent repair processes.

Delayed healing or fracture nonunion can be explored by creating a certain fracture gap (segmental bone defect/critical size defect) which is later bridged and stabilized by plates and screws or an external fixation device. Stripping of the adjacent periosteum further compromises healing which reproducibly results in nonunion.

Open/Closed Fractures – Fractures can be achieved by applying different techniques such as a manual guillotine or impact device. The closed fracture model using a guillotine introduced by Bonnarens and Einhorn is most commonly used [11]. This model can quite reproducibly imitate secondary fracture healing. Fracture stabilization can be achieved by introducing an intramedullary rod either prior or after using the guillotine.

Open techniques allow precise localization of the fracture site but implicating the disadvantage of washing out the initial fracture hematoma. The soft tissue envelope may sufficiently stabilize the fracture site even without fixation, whereas the open approach necessitates fixation methods. Additionally, surgery accompanied with risks and variables (e.g., damage to the surrounding soft tissue, risk of infection) could potentially influence fracture healing outcome.

■ **Fixation Techniques**

Internal endogenous fracture stabilization can be practiced in the radial/ulnar model, the radius being fractured and the ulna functioning as the strut to maintain length, rotation, and alignment.

Intramedullary stabilization through nailing is a relatively easy approach and is applicable in animals of all sizes. It allows investigating indirect fracture repair and treatment modalities to augment the healing process. The rod can be introduced prior or after the fracture has been created. Unrestricted postoperative rehabilitation or movement is possible while the nail preserves alignment of the bony fragments. However, the diameter of the nail has to be chosen carefully, because it was shown that smaller diameters in relation to the medullary cavity will result in nonunion [12]. Biomechanical testing can be performed easily by removing the nail distant to the fracture site without influencing the local healing response.

Plating and screw fixation are valuable methods in studying direct fracture repair. Surgery can be quite demanding when a small-scaled animal model, i.e., mice, is used. As mentioned, direct visualization facilitates the fracturing process but bears also disadvantages which have to be taken into account. Radiological assessment of the fracture is commonly used to judge osseous healing, but plates can interfere with radiological evaluation. Biomechanical testing is yet another method to evaluate callus/bone quality at the fracture site. This can only be realized when plates are removed prior to euthanasia causing additional iatrogenic trauma potentially interfering with subsequent testing.

External fixation, which can either be unilateral or circular, has several advantages including the relative ease of application distant from the fracture site as well as fracture execution in a closed manner without additionally harming the soft tissue coverage by the surgical intervention. Moreover, radiological evaluation is less complicated, removal is easy, and interference with postmortem assessments is avoided.

Finally, *splints or casts* are an additional method to immobilize the fractured hind limb. However, these fixations are often badly tolerated, chewed on, or soiled by the animals.

7.6 Physiology of Fracture Healing

In the context of bone regeneration, the following principles are essential:
- *Osteoconduction*: It is passive process that involves adhesion and ingrowth of cells due to local material properties such as porosity and surface roughness.
- *Osteoinduction*: It is an active process that involves recruitment and differentiation of cells. Growth factors which are able to induce ectopic bone formation, e.g., BMP-2, BMP-7, TGF-β1, IGF-1 (derived from bone), PDGF, EGF, VEGF, and TGF-β1 (from platelets).
- *Osteogenesis*: Cells are called osteogenic, if they possess the potential to differentiate into active bone-forming osteoblasts.

Looking closely at the fundamentals of fracture consolidation, we have to differentiate between primary (intramembranous) and secondary (endosteal/intramembranous) healing.

7.6.1 Primary/Direct (Intramembranous) Healing

Primary fracture healing implies absolute stability, consisting in anatomic reduction and compression restricting any kind of motion at the fracture site. Therefore this type of healing can only be seen in cases of ORIF yielding absolute stability. In these cases direct remodeling of lamellar bone, of the Haversian canals, and of blood vessels occurs. Primary or direct fracture healing happens either through contact or gap healing.

Contact Healing

Direct bone healing as mentioned above can only occur when an anatomic reduction is achieved and rigid fixation is provided, resulting in a substantial decrease in interfragmentary strain. Due to the anatomic reduction, fracture ends are in direct contact reestablishing cortical continuity.

If the fracture gap is less than 10 µm and the interfragmentary strain is less than 2%, fracture healing occurs through contact healing [13].

Under these conditions, so-called cutting cones are formed at the ends of the osteons next to the fracture [14]. These cutting cones gradually cross the fracture line. Osteoclasts sit at the convex side (tip) of the cutting cones, acting as a bur in order to create longitudinal cavities at a rate of around 50–100 µm/day. These cavities (tunnels) allow penetration of blood vessels, endothelial cells, and osteoprogenitor cells, which further differentiate into bone-forming osteoblasts at the rear end of the cones. Due to the coupling of bone resorption and bone formation, bony union and simultaneously restoration of the Haversian system are achieved in an axial direction.

The osteons bridging the fracture gap later mature directly into lamellar bone avoiding the formation of woven bone as intermediate step, as it is the case in secondary fracture healing. This results in direct fracture healing, which per definition circumvents periosteal callus formation.

Gap Healing

In contrast to contact healing, where bony union and Haversian remodeling occur simultaneously, in gap healing these steps take place successively.

As in contact healing, gap healing occurs if stable conditions and anatomic reduction are achieved, although the fracture gap can be somewhat bigger but must be less than 800 µm to 1 mm.

Initially the fracture site is filled by lamellar bone, oriented perpendicular to the long axis, requiring a secondary osteonal reconstruction unlike the process of contact healing. The primary bone structure is then gradually replaced by longitudinal revascularized osteons carrying osteoprogenitor cells, which differentiate into osteoblasts and produce lamellar bone temporarily stabilizing the fracture. Initially this lamellar bone is mechanically weak as it is laid down perpendicular to the long axis of the bone. After 3–8 weeks secondary remodeling takes place. This process resembles the one seen in contact healing and leads to the consolidation of bone [13].

7.6.2 Secondary/Indirect (Endosteal/Intramembranous) Healing

Secondary or indirect fracture healing is the most common form of fracture healing, consisting of both endochondral and intramembranous bone healing. Micromotion and weight-bearing enhance callus formation; therefore, anatomic reduction or rigid fixation, a principle used in compression plating, is not necessitated. However, delayed osseous healing or nonunion may occur if micromotion exceeds a certain level. Indirect fracture healing can be seen in conservative fracture treatment such as bracing and casting as well as in certain operative treatments. These include all types of treatments yielding a so-called relative stability according to the AO principles of fracture care like intramedullary nailing and external or internal fixation of complicated comminuted fractures along with bridging plating as applied in minimally invasive plate osteosynthesis (MIPO).

Secondary or indirect fracture healing involves a combination of intramembranous and endochondral ossification. According to Einhorn there are five distinct stages of healing. These include an initial phase in which a hematoma is formed and inflammation occurs, followed by a subsequent stage of angiogenesis and initial cartilage formation. Thereafter, three successive processes of cartilage calcification, cartilage removal, and bone formation and ultimately a more chronic stage of bone remodeling occur. The intramembranous bone is formed by committed osteoprogenitor cells that reside in the cambium layer of the periosteum [15]. As mentioned in the introduction, fracture healing follows a strict time course in order to allow osseous consolidation.

Inflammation

Initial hematoma formation after trauma is of paramount importance. Upon activation of the clotting cascade, the hematoma partially acts as glue in order to limit motion at the fracture ends. Additionally it provides a template for callus formation which occurs in two phases comprising a soft (fibrous/cartilaginous) and later a hard (osseous) callus. Blood cells are recruited and an acute inflammatory response is initiated in order to propagate healing. Secretion of TNF-α, interleukin-1 (IL-1), interleukin-6 (IL-6), IL-11, and IL-18 leads to further recruitment of inflammatory cells as well as to the promotion of angiogenesis [16, 17]. In vitro TNF-α induces osteogenic differentiation of mesenchymal stem cells (MSCs), mediated by activation of its two receptors TNFR1 and TNFR2 being expressed on both osteoblasts and osteoclasts. Since TNFR1 is constantly expressed in bone and TNFR2 expression is only observed following injury, a more specific role of TNFR2 in bone regeneration is to be presumed.

Within the interleukin family, IL-1 and IL-6 are considered to be the most important for fracture healing.

IL-1 expression, produced by macrophages in the acute phase of inflammation, overlaps with that of TNF-α in a biphasic mode and induces production of IL-6 in osteoblasts. Furthermore, it promotes the production of the primary cartilaginous callus and angiogenesis at the injured site, doing so by activating either of its two receptors, IL-1RI or IL-1RII [18].

IL-6, being an acute phase protein, stimulates angiogenesis, vascular endothelial growth factor (VEGF) production, and differentiation of osteoblasts as well as osteoclasts.

Repair

Fracture healing or bone regeneration is dependent on recruitment, proliferation, and differentiation of mesenchymal stem cells (MSCs), which are mobilized locally (surrounding soft tissues and bone marrow) and systemically. To knowledge it appears that recruitment and homing of these MSCs is mediated by bone morphogenetic proteins (BMP-2, BMP-7) and stromal cell-derived factor-1 (SDF-1) as well as its corresponding receptor CXCR-4, being a G-protein-coupled chemokine receptor.

After the primary hematoma has formed, a fibrin-rich granulation tissue develops. Within this tissue, endochondral bone formation in between fracture ends and external to periosteal sites can be observed, whereas subperiosteally intramembranous bone formation occurs. The regions of endochondral ossification are less mechanically stable, and the cartilaginous tissue forms a soft callus acting as scaffold, giving the fracture a somewhat stable structure [19].

In animal models (rat, rabbit, mouse), the peak of this soft callus formation occurs 7–9 days post trauma [15].

Cho et al. have shown that TGF-β_2, TGF-β_3, and GDF-5 are involved in chondrogenesis and endochondral ossification underlining the importance of the TGF-β superfamily in fracture healing [20]. According to Marsell and Einhorn, cell proliferation in intramembranous ossification at periosteal sites is induced by BMP-5 and BMP-6.

Furthermore, as stated above, BMP-2 has been proven to be crucial for initiation of the healing cascade. This concept was verified by Tsuji et al. using BMP-2 knockout mice, which were not able to form callus to heal their fractures successfully.

For bone regeneration to progress, the soft cartilaginous callus initially built needs to be resorbed and substituted by hard (bony) callus.

To some extent, this step in the fracture healing cascade recapitulates embryological bone development. Cellular volume and matrix deposition are increased through a combination of cellular proliferation and differentiation.

The differentiation of pluripotent MSCs into the osteoblastic lineage is thought to be regulated by the Wnt family. This signaling pathway has been shown to be a very important complementary pathway of RANK and its ligand receptor activator of nuclear factor kappa B ligand (RANKL) in cell differentiation as well as bone formation. Furthermore, the Wnt pathway signaling through β-catenin seems to positively regulate osteoblastic bone formation at later stages of development [21].

The extracellular matrix becomes calcified as fracture callus chondrocytes proliferate and turn hypertrophic. This cascade is primarily coordinated by RANKL, macrophage colony-stimulating factor (M-CSF), osteoprotegerin (OPG), and TNF-α, initiating resorption of this newly mineralized cartilage [16, 22]. It is thought that throughout this process, M-CSF, RANKL, and OPG help to recruit bone cells and osteoclasts to form woven bone.

Even though the most important role of TNF-α may be to initiate chondrocyte apoptosis, it also promotes the recruitment of MSCs with osteogenic potential (osteogenesis) [16].

Mitochondria accumulate calcium-containing granules created in the hypoxic fracture environment and therefore play an important role in the calcification mechanism.

After emigration of fracture callus chondrocytes into the cytoplasm, calcium granules are carried into the extracellular matrix, precipitating with phosphate forming initial mineral deposits which finally build up apatite crystals.

In animal models the peak of the hard callus formation is usually reached by day 14. This can not only be seen by histomorphometry of the mineralized tissue but also by expression of extracellular matrix markers such as type I procollagen, osteocalcin, osteonectin, and alkaline phosphatase [15]. As the hard callus formation advances, the calcified cartilage is gradually replaced by woven bone, making the callus stronger and more mechanically rigid.

Remodeling

The hard callus as rigid structure provides some biomechanical stability. The former biomechanical properties of normal bone however are still to be reached.

In order to reestablish the biomechanical properties of bone prior to fracture in the sense of a true regeneration, a second resorptive phase is initiated, and the hard callus is remodeled into a lamellar bone structure with a central medullary cavity [16].

IL-1 and TNF-α biomechanically coordinate this phase, showing high expression levels during this stage. In contrast, most members of the TGF-β family already show diminished expression at this later phase.

Some BMPs such as BMP-2 show continuous moderately high expression levels and therefore apparently also stay involved in this phase.

During remodeling a constant balance is kept between callus resorption on the one hand and lamellar bone formation on the other hand. Remodeling starts as early as 3–4 weeks after fracture, but may take years to be completed in order to regain the above-mentioned former biomechanical properties of bone.

Overall this process is age dependent and thus occurs faster in younger patients and animals. Mechanical loading also influences remodeling through creation of electrical polarity. Axial loading of long bones on the one hand creates electropositivity at the convex side activating osteoclasts and on the other hand electronegativity on the concave side alternatively activating osteoblasts. In doing so the external callus is gradually replaced by lamellar bone formed by osteoblasts, whereas the internal callus is resorbed by osteoclasts reestablishing the medullar cavity inherent to the normal structure of diaphyseal bone [23].

Cell types involved in bone regeneration	
«Bone lining cells»	Flat inactive cells that reside at the outer surface of bone as well as in Haversian canals. They are spindle shaped and possess only few amounts of roughened endoplasmatic reticulum (RER) and Golgi complexes, reflecting their inactive metabolic state
Osteo-blasts	Active cells that produce collagens, proteoglycans, as well as glycoproteins (osteocalcin, bone sialoprotein, alkaline phosphatase, etc.) residing on the surface of bone trabeculae. Osteoblasts are connected to one another via thin cytoplasmatic processes. In contrast to bone lining cells, they show all histological signs of active protein-producing cells such as high amounts of RER. Consistently they are basophil. The newly formed organic extracellular matrix (ECM) is named osteoid. In the further course, osteoid is mineralized. Osteoblasts produce around 1 μm of osteoid per day, 70% of which is mineralized within the first few days. As soon as osteoblasts are fully embedded in their own ECM, they become osteocytes
Osteocytes	Osteocytes are connected through thin filopodial processes which run in canaliculi. Gap junctions enable ions and molecules to travel across up to 15 cells. Due to their interconnectivity, osteocytes are thought to play an important role in mechanotransduction
Osteoclasts	In contrast to all of the abovementioned cells, which derive from mesenchymal stem cells, osteoclasts derive from myeloid precursors of the hematopoietic lineage. They are a fusion product of mononuclear cells (syncytium) and therefore contain multiple nuclei. Active osteoclasts settle in so-called Howship lacunae. The resorbing surface of the osteoclast is enlarged by a ruffled border. Osteoclasts produce a significant amount of tartrate-resistant acidic phosphatase (TRAP) for demineralization of bone as well as cathepsin K (a cysteine protease) for degradation of collagen type I and other noncollagenous proteins

Fracture Healing

Key factors in bone repair

Key molecule	Function	Effects	Expression pattern
Extracellular factors			
IL-1, IL-6, TNFα ILx – interleukin-x TNF – tumor necrosis factor α	Elicit inflammation and migration IL-1 – secretion of IL-6, GMCSF, MSCF IL-6 – very sensitive to IL-1	In vitro inhibit osteoblastic differentiation, but in vivo TNFα is crucial for bone repair; role of IL-6 is controversial (anti- or pro-osteogenic probably, depending on soluble IL-6 receptor)	IL-1 – produced by macrophages in a temporal manner throughout fracture healing being highest in the early inflammatory phase IL-6 – produced by osteoblasts in a temporal manner
SDF1 Stromal cell-derived growth factor	Chemotactic factor	Homing of stem cells and recruitment of inflammatory cells	Released in response to hypoxic harmed local tissue; endogenous CXCR4 expressing stem cells are recruited
TGFβ Transforming growth factor β	Chemoattractant for macrophages, mitogenic factor, osteogenic factor, differentiation factor of periosteal progenitor cells, promotes angiogenesis	Can induce osteoblast differentiation at the early stage of immature cells but can also inhibit osteogenesis in committed cells	– Produced by platelets, inflammatory cells, osteoblasts, osteoclasts, mesenchymal cells, endothelial cells, and periosteal mesenchymal cell chondrocytes –Weakly expressed in proliferating mesenchymal cells and endothelial cells in the inflammatory phase, strongly expressed in proliferating osteoblasts during intramembranous ossification, and strongly expressed in proliferating chondrocytes during chondrogenesis and endochondral ossification
BMP-2, BMP-4, BMP-7 Bone morphogenetic protein-x	Osteogenic factors	2 – might initiate bone formation healing, induce other BMPs 4 – stimulates TGF-b, binds to collagen I and IV 7 – induces Osf2/Cbfa1 (transcription factor for early osteoblast differentiation)	Produced by mesenchymal and osteoprogenitor cells, fibroblasts, and proliferating chondrocytes; strong expression during inflammation (undifferentiated mesenchymal cells) and intramembranous ossification (proliferating osteoblasts)

(continued)

Key factors in bone repair (continued)			
Key molecule	Function	Effects	Expression pattern
Noggin	BMP-2, BMP-4, and BMP-7 specific inhibitor	Suppresses osteoblastic differentiation	Coordinated and similar expression as BMP4, suggesting important role of noggin/BMP4 balance in regulation of callus formation
FGFb Fibroblastic growth factor	Promotes angiogenesis and is a mitogenic factor; osteogenic factor (controversially discussed)	Can stimulate Sox9; might be a negative regulator of postnatal bone growth and remodeling	Expressed throughout fracture healing (macrophages – inflammation; osteoblasts – intramembranous ossification; chondrocytes – chondrogenesis; osteoblast/chondrocytes – endochondral ossification)
IGF-I, IGF-II Insulin-like growth factor	Mitogenic factors, osteogenic factors	Increase collagen syntheses, inhibit collagen degradation; proliferation of osteoblasts; increased osteoclast formation (I) and modulation of function (II)	I – during intramembranous ossification in osteoblasts; mRNA peak at day 8 after fracture II – proliferating chondrocytes; largely negative in osteoblast during remodeling
VEGF Vascular endothelial growth factor	Angiogenic and vasculogenic factor	Most potent pro-angiogenic and vasculogenic factor; crucial at the onset of bone formation; osteoblast and osteocyte differentiation, osteoclast recruitment	During endochondral and intramembranous ossification
PlGF Placental growth factor	Angiogenesis, vasculogenesis (member of the VEGF family)	Induces proliferation and osteogenic differentiation of MSCs Essential for vascularization	Expressed throughout repair
PDGF Platelet-derived growth factor	Mitogenic and chemotactic factor	Highly mitogenic factor for MSCs, connective tissue cells; initiates callus; chemotactic for MSCs, osteoblasts, and perivascular cells	Weakly expressed during inflammation; constantly in the callus by platelets, macrophages, monocytes, and endothelial cells

Fracture Healing

Key factors in bone repair (continued)

Key molecule	Function	Effects	Expression pattern
Wnts Wingless-type MMTV integration site family	Mitogenic and osteogenic factors	Depending on Wnt type, crucial for osteoprogenitor proliferation; can also inhibit final osteoblast maturation	Upregulated in early fracture healing, declines to slightly suprabasal levels 14–21 days post fracture
DKK1 Dickkopf	Inhibitor of Wnt signaling	Strongly inhibits osteogenesis of MSC and osteoprogenitor cells; can stimulate terminal maturation	
Ihh Indian hedgehog	Osteochondrogenic factor	Pivotal role for growth plate and endochondral formation; can inhibit osteoblast differentiation; might induce PTHrP expression	Expressed in (pre-)hypertrophic chondrocytes (1–2 weeks post fracturing) and in osteoblasts close to endochondral ossification front (3rd week)
PTHrP Parathyroid hormone	Autocrine and paracrine osteochondrogenic factor, anabolic	Pivotal role for growth plate and endochondral formation; can induce or inhibit osteogenesis	PTHrP mRNA expression in preosteoblasts, differentiated osteoblasts, and osteocytes during intramembranous ossification. In the bony calluses, both PTHrP mRNA and its receptor were detected in the osteoblasts and the osteocytes in the trabecular bones near the fracture sites
OPG Osteoprotegerin	Decoy receptor of RANKL, inhibition of RANKL	Strongly inhibits bone resorption and has a pivotal role in bone remodeling	Constitutively expressed in unfractured bones and elevated levels throughout the repair process (peak at 24 h and 7 days)
RANKL	Induces osteoclastogenesis	Strongly stimulates bone resorption and has a pivotal role in bone remodeling	Expression is minimal in unfractured bones but strongly elevated throughout the period of fracture healing (peak at 3 and 14 days post fracture)
M-CSF	Induces osteoclastogenesis	Crucial for osteoclastogenesis.	Follows temporal expression patterns of RANKL
Gastrointestinal serotonin	Neurotransmitter inhibiting osteogenesis	Inhibits bone formation and repressed by Lrp5	Expressed by enterochromatin cells

(continued)

Key factors in bone repair (continued)			
Key molecule	Function	Effects	Expression pattern
Intracellular messengers			
MAPKs Mitogen-activated protein kinases	Transduce osteogenic signaling by phosphorylation	Crucial for regulation of intracellular signaling induced by osteogenic factors (still controversial)	Throughout the fracture healing process
β-Catenin	Osteogenic transducer factor	Pivotal role in transducing osteogenic signal from Wnt and is negatively regulated by GSK3b	In early stages, beta-catenin is required for pluripotent mesenchymal cells (differentiation to osteoblasts or chondrocytes); positively regulates committed osteoblasts
Runx2/Cbfa1 Runt-related transcription factor 2/core-binding factor subunit alpha-1	Early osteogenic transcription factor (master osteoblast transcription factor)	Master regulator of early osteogenesis; osteoblastogenesis and terminal chondrocyte maturation	Involved in both the intramembranous and endochondral ossifications
Osx (Sp7) Osterix	Late osteogenic transcription factor	Master regulator of late osteogenesis, inhibiting chondrogenesis	Runx2-induced transcription factor expressed in osteogenic cell progenitors stimulating osteoblastic differentiation rather than chondroblastic
NF-kB	Inflammation transducer factor, inhibits osteogenesis	Inhibits the differentiation of MSCs, and committed osteoblastic cell inhibition will not only suppress osteoclast-mediated bone resorption but will also promote osteoblast function and bone formation	Early bone fracture healing by endochondral ossification depends on a hematoma-induced inflammatory environment, and several NF-κB-target genes (e.g., IL-6, TNFα, COX2) are involved in bone fracture repair; chondrogenic differentiation is facilitated by early transient activation of NF-κB/p65
Adapted from Deschaseaux [24] and Sfeir [25]			

7.7 Pathophysiology of Fracture Healing

It is estimated that delayed or impaired healing will occur in 5–10% of the 5.6 million fractures that occur annually in the United States, and up to 10% of all fractures will require additional surgical procedures for impaired healing [26].

Disturbances in physiological bone healing potentially result in delayed healing or nonunion of fractures. Although nonunions represent an infrequent entity (approximately 2.5% of all fractures in Austria develop a nonunion), they constitute a highly relevant medical as well as socioeconomic issue. Unfortunately, a unified classification regarding delayed healing and nonunions is still lacking. thus making cross comparisons difficult.

In most cases, normal fracture healing occurs within 3 months. Therefore, according to most textbooks, delayed healing is estimated to take place between 3 and 6 months. Consequently, if a fracture fails to show any signs of osseous consolidation after 6 months, it can be defined as nonunion. The Federal Food and Drug Administration (FDA), however, defines nonunion as a fracture which does not heal on its own within 9 months, showing no radiological signs toward osseous healing over the past 3 months.

Factors contributing to delayed healing or nonunion development can be categorized into local (patient independent) and systemic (partially patient dependent) factors. Local factors include the anatomic region and type of fracture as well as vascularization and soft tissue damage associated with the initial trauma. Hence, the tibia is more prone in developing healing disturbances due to its scarce soft tissue coverage and high probability of open and comminuted fractures.

Systemic factors, which can be partly influenced by the patient, include comorbidities such as diabetes, atherosclerosis, congestive heart disease, metabolic syndrome, lifestyle habits such as smoking and alcohol consumption, as well as current medication and patient's age.

7.7.1 Classifications of Nonunions (NU)

Based on different criteria, there are several classification systems. Among these the most commonly used is the classification according to Weber and Cech [27], which roughly differentiates between vital and avital nonunions. Vital nonunions are further subdivided into hypertrophic, normo−/oligotrophic nonunions. These nonunions, also referred to as reactive nonunions, generally have the potential to heal with appropriate measures such as immobilization. In contrast, avital or nonreactive nonunions, farther subdivided into dystrophic, atrophic, necrotic/avascular, and segmental defect nonunions, show distinct disturbed healing without having the ability to heal on their own.

Besides the Weber and Cech classification, radiological assessment of nonunions separates them into hypertrophic, oligotrophic, and atrophic. Excessive callus formation or hypertrophic nonunion, also named elephant foot, is mainly seen due to inadequate initial fracture stability. Although not yet united, the bone is vital and shows sufficient perfusion thus bearing the inherent ability to heal, after adequate stabilization is performed.

In contrast, oligotrophic nonunions show only little to no callus formation on plain radiographic films. Nevertheless, as mentioned above, they are categorized as being biologically active and thus still have the potential to heal.

A local factor which plays a major role in the fate of fracture healing is local blood supply. Perfusion can be compromised either through the initial trauma or iatrogenic due to an extensive surgical approach. On X-ray this can be seen as atrophic nonunion.

Another important factor potentially interfering with the fracture healing process is infection. If present, infection leads to the development of septic nonunion. In most cases staphylococci (S. aureus and S. epidermidis) act as the main pathogen. These bacteria either enter the wound at the initial trauma (open fracture) or during the further cause of intervention, for instance, the operation. Primary open fractures are significantly more prone to develop septic nonunions. In order to prevent the bone from developing an acute, subacute, or chronic osteitis/osteomyelitis, septic nonunions have to be recognized and treated in a timely manner. Serological testing should include erythrocyte sedimentation rate (ESR), total leucocyte count and relative number of segmented neutrophils, C-reactive protein (CRP), interleukin 6 (IL-6), and procalcitonin. All these parameters together show a high sensitivity and specificity in detecting infections, even though single parameters can be negative. Especially in chronic or low-grade infections, false-negative results (low sensitivity) are possible, making diagnosis more difficult. In conclusion, blood testing shows a high negative predictive value. Therefore negative results can help to rule out infection, whereas positive results are not evidentiary for infectious disease.

7.7.2 Molecular Alterations

Inflammation is an essential process not only for normal fracture repair but for tissue healing in general. However, following the initial inflammatory reaction, anti-inflammatory stimuli are necessary to proceed with the physiological healing cascade. An excessive and/or prolonged inflammatory response ultimately leads to complications such as delayed bone healing or nonunion. It could be shown that the adaptive and innate immune system oppositionally contributes to bone repair. While the adaptive immune system has a negative effect on healing, the innate immunity partially has positive effects on repair. The latter could be shown by blocking early macrophage infiltration through chemokine (C-C motif) receptor 2 inactivation, resulting in delayed fracture repair due to impaired function of osteoclasts.

Recently, bone-resident macrophages (osteomacs) have been identified, which were shown to influence fracture healing and bone remodeling. In vivo depletion of these cells resulted in delayed fracture healing, whereas stimulating migration of these cells to the fracture site by macrophage colony-stimulating factor 1 (CSF-1) leads to collagen type I deposition and mineralization. These findings suggest a fundamental role of this macrophage subpopulation in anabolic activities during fracture healing.

Comorbidities such as diabetes, smoking, or aging show increased inflammation associated with a decreased endogenous potential of bone healing. Aged mice were transplanted with juvenile bone marrow in order to evaluate the ability to stimulate fracture healing. In fact, this approach exhibited an enhanced healing combined with faster resolution of inflammation. Similarly, inhibiting macrophage migration during fracture repair in old animals demonstrated superior healing outcome in comparison to an age-matched control group.

The age dependency seen in the inflammatory response during fracture repair is potentially caused by a reduced ability to switch from a pro- to an anti-inflammatory status.

Accelerated healing of calvarial defects could be observed in a transgenic mouse model lacking the Toll-like receptor 4, supporting the thesis that the innate immune system may negatively modulate fracture healing.

Prostaglandins are other key inflammatory mediators influencing osteoclasts as well as osteoblasts to promote bone regeneration. Cyclooxygenase 2 (COX-2) – an ubiquitous enzyme catalyzing prostaglandin synthesis – is markedly upregulated after bone injury and impairs fracture healing if selectively inhibited. Osteogenesis is directly triggered by COX-2 via induction of cbfa-1 (core-binding factor subunit alpha 1) and osterix (2 downstream genes of COX-2) which stimulate the differentiation of MSCs into functional osteoblasts leading to intramembranous and endochondral ossification. An additional important source of stem and progenitor cells is the periosteum, where COX-2 exhibited a crucial role for the activation and differentiation of these cells after bone injury.

Due to the abovementioned influences on bone healing of selective and nonselective COX-2 inhibitors, the use of nonsteroidal anti-inflammatory drugs (NSAIDs) as pain medication should be avoided in patients suffering from fractures.

Yet another key player in bone regeneration is angiogenesis. Damage to the local blood supply has been shown to dramatically enhance the chance to develop a nonunion. Regulation of angiogenesis during bone regeneration is orchestrated by a variety of growth factors, cytokines, adhesion molecules, and enzymes degrading the extracellular matrix. Exemplarily, the initial angiogenic response following a fracture is impaired when vascular endothelial growth factor or placental growth factor is lacking, resulting in reduced ossification and delayed fracture healing.

Hypoxia or ischemia at the fracture site effectively upregulates hypoxia-inducible factor 1α, which is a key transcription factor regulating multiple downstream genes associated with angiogenesis. Genetically manipulated osteoblasts overexpressing hypoxia-inducible factor 1α enhances not only angiogenesis but also bone repair.

In the time course of bone healing, angiogenesis promotes the resorption of cartilage, which is essential to induce osteogenesis. Matrix metalloproteinases (e.g., MMP-9) degrade the cartilaginous matrix, continuously releasing the pro-angiogenic VEGF resulting in neovascularization and ossification.

Compromised angiogenesis can be found in the elderly with disturbed healing which underlines the essential role of angiogenesis in bone regeneration.

Regular fracture healing involves numerous molecules and signaling pathways; thus, proper regulation of these factors is indispensable. The transforming growth factor beta superfamily contains signaling proteins such as bone morphogenetic proteins which have strong osteoinductive properties. Endothelial cells also express bone morphogenetic proteins underlining the supportive role of angiogenesis in bone formation.

The bone morphogenetic protein pathway interacts with a variety of other pathways which are crucial in bone homeostasis. One of these pathways is the Wnt signaling pathway, which is not only involved in tissue repair including bone but also in embryogenesis and organogenesis. If Wnt signaling is altered fracture healing is affected. Inactivation of the Wnt co-receptor low-density lipoprotein receptor-related protein 5 (LRP5) negatively influences fracture repair. Binding of Wnt to LRP5 leads to nuclear translocation of β-catenin which is the initial step of a canonical signaling pathway influencing many cell types involved in bone regeneration including periosteal stem cells. β-Catenin exhibits differential function depending on the stage of fracture healing. In the early stages, it is essential to trigger differentiation of MSCs to chondrocytes and osteoblasts [21]. During the maturation of osteoblasts to osteocytes, β-catenin decreases thus highlighting that higher β-catenin levels potentially enhance bone formation.

Among alterations in these signaling pathways, the hormonal dysregulation of parathyroid hormone seems to substantially impair fracture healing on multiple steps of the repair process. Both the cartilaginous and the bony callus formation are less pronounced in consequence of downregulated osteoblastic gene and protein expression due to endogenous parathyroid hormone (PTH) deficiency [28]. Intermittent parathyroid hormone is applied in osteoporotic patients intending to increase bone formation being the only

FDA-approved osteo-anabolic drug. It has been shown that intermittent PTH augments maturation of osteoblasts while reducing sclerostin, an antagonist of Wnt signaling being almost exclusively expressed by osteoblasts and osteocytes. Hence, it is likely that PTH and Wnt promote bone formation through complementary pathways. At the fracture site, intermittent PTH upregulates Wnt, thus increasing bone formation and stimulation of bone remodeling.

7.7.3 Treatment Modalities

The existence of delayed osseous healing or nonunion has to be carefully evaluated prior to any further intervention. Several techniques are available. Recently, systemic and local biomarkers were clinically introduced to assess and/or predict nonunions. However, to date, no reliable predictive biomarker was identified which would allow to make an informed clinical decision on the subsequent treatment. Therefore, the clinical assessment (e.g., pain during weight-bearing, swelling) in conjunction with radiological examination (conventional X-ray, computed tomography, nuclear medicine) still represents the standard procedure to judge nonunions.

The aim of any therapeutic intervention should be the bony consolidation along with pain-free weight-bearing and satisfactory functionality. Recognizing the complex process of fracture repair, the diamond concept was introduced by Giannoudis et al. [29] drawing attention to key factors necessary for regular healing. Originally derived from the tissue-engineering approach with a triangular concept comprising growth factors (osteoinduction), scaffolds (osteoconduction), and mesenchymal stem cells (osteogenesis), the significance of the mechanical environment (mechanical stability) was amended to the concept.

Furthermore, any therapeutic approach has to be oriented based on the respective pathology. In most cases, hypertrophic nonunions are the result of insufficient initial stability. Introducing additional stability with a more rigid fixation would suffice as treatment in most of these cases. In atrophic nonunions, however, the fracture ends are avital therefore necessitating further treatments enhancing perfusion and osteogenesis.

- **Mechanical Stability**

Depending on fracture localization and type of nonunion, the appropriate osteosynthetic material has to be chosen. In any case, the soft tissue envelope has to be treated carefully to avoid further compromise of the microenvironment. Periosteal stripping must be avoided as it was shown in experimental models that deprivation of the blood supply reproducibly leads to nonunions. Resection of the interposed fibrous scar tissue as well as avital bone in the nonunion gap is essential for induction of bleeding, enabling stem cell recruitment. Additionally correct fracture reduction along with compression should be achieved. In selected cases of delayed fracture healing, dynamization of an intramedullary nail might be sufficient in order to allow compression and consequently fracture healing. Another option is reamed compression nailing with exchange of the initial intramedullary device in order to propagate proper healing. Reaming is supposed to induce endosteal bleeding with subsequent recruitment of stem cells as well as reversing endosteal to periosteal perfusion.

In atrophic nonunions plating (sometimes even as double plating) in combination with osteoinductive factors as well as osteoconductive materials (e.g., autologous bone graft) can be used as treatment. Locking plates might be beneficial in these occasions.

■ Autologous Bone Graft

Despite major advances in tissue engineering, autologous bone grafting (ABG) remains the gold standard in the treatment of nonunions. The first documented case of an autologous bone grafting procedure dates back to 1668. The main advantage of autologous bone grafting (ABG) compared to other treatment modalities is the combination of osteoinductive (e.g., BMPs), osteoconductive (e.g., collagen and hydroxyapatite), and osteogenic (stem cells) properties in a single procedure. Bone grafts can be divided into cortical, cancellous, and corticocancellous grafts. Cortical grafts are primarily used as strut grafts in order to support the structural network. Cancellous grafts on the other hand are the main source of stem cells, combining osteogenic and osteoconductive properties, whereas corticocancellous grafts obviously combine both of the aforementioned. Frequently used sources of grafts include the ilium (anterior and posterior aspect), tibia (anteromedial), or fibula. While grafts originating from the ilium and the tibia can be either cancellous, cortical, or both, the fibula can be harvested as vascularized or nonvascularized full transplant. Despite indisputable advantages of ABG, there are two crucial negative aspects which have to be considered. Firstly, ABG requires an additional surgical procedure inherently causing complications such as donor site morbidity mainly comprising pain, infection, and hematoma formation. Secondly the amount of autologous bone grafts, which can be harvested, is limited.

■ Bone Grafts of Different Origin

The body's bone is the most preferred source for grafts, lacking risk of graft rejection. However, under some circumstances (e.g., multiple surgical revisions), ABG might not be applicable or indicated, thus necessitating alternatives. In these cases, allografts, xenografts (heterogeneous grafts), or bone substitutes can be considered. Allografts are usually taken from deceased humans who donated their bone. This bone is then prepared and stored by a certified tissue/bone bank. Allografts can be differently processed and purchased as one of the following:
1. Fresh or fresh-frozen bone
2. Freeze-dried bone allograft (FDBA)
3. Demineralized freeze-dried bone allograft (DFDBA)

Heterogeneous grafts or xenografts originate from other species than human and were used in the past, but recommendation was denied, because of infection and rejection issues as well as unsatisfactory outcome in general.

Bone substitutes nowadays mainly consisting of calcium phosphates, e.g., hydroxyapatite or tricalcium phosphate, primarily act as scaffolds for migrating osteoprogenitor cells before being replaced by endogenous bone. Depending on their origin, being natural or synthetic, they can be fully or partially resorbed. Another representative of this group is calcifying marine algae, such as *Corallina officinalis*, which resembles cancellous/spongious bone and is gradually resorbed in the later course.

- **Reamer Irrigation Aspiration (RIA System)**

The RIA system allows retrieval of 25–90 cm^3 intramedullary bone graft during reaming of the medullar cavity. The obtained material contains not only endogenous stem cells but also osteoinductive growth factors facilitating nonunion treatment.

- **Bone Marrow Aspiration**

Stem cells can be quite easily obtained by percutaneous bone marrow aspiration. However, a wide variability in the cell quantity can be found which can be improved by a specific purification technique. The percutaneous approach minimizes donor site morbidity as seen in autologous bone graft harvesting, making this modality especially valuable in small bony defects where extensive concomitant soft tissue damage is present.

- **Growth Factors**

Over the past a couple of different osteoinductive molecules have been tested experimentally and clinically. Among these, BMP exhibited the most osteoinductive potential inducing the differentiation of residential mesenchymal stem cells into bone-forming osteoblasts. So far the clinical application includes the recombinant human isoforms BMP-2 (Infuse®/Medtronic) and BMP-7 (OP-1®/Stryker), the latter being approved for traumatic recalcitrant tibial nonunions. Friedlaender et al. compared the clinical and radiological efficacy of BMP-7 with autologous bone grafting and found similar results avoiding donor site morbidity in the BMP-7 group. Furthermore, the combination of both BMP-7 and autologous bone grafting showed synergistic effects.

- **Noninvasive Biophysical Stimulation**

Extracorporeal Shockwave Therapy (ESWT)

Recently, ESWT has emerged as a valuable approach in the treatment of nonunions capable to avoid complex surgical procedures. The noninvasive character and the cost-saving aspect are key advantages over other treatment options. In 1998 the AUVA Trauma Center Meidling, Vienna, Austria, was one of the first hospitals worldwide using ESWT as a standard procedure for nonunion treatment. Since then over 2500 patients were treated successfully using different electrohydraulic shockwave devices. Overall, we could achieve bony union in approximately 80%, the tibia responding best with a healing outcome of up to 90% clinically and radiologically assessed 6 months after treatment. Besides some temporary local redness, minor local swellings, or petechial bleeding in some of the patients, so far no serious treatment-associated complications were observed. Based on our data and recent literature, ESWT can be recommended as the therapy of first choice for nonunions. Advantages include the high efficacy, lack of complications due to avoidance of surgical interventions potentially associated with further complications, higher patient convenience, and considerable lower costs.

Shockwaves are acoustic waves propagating in medium at supersonic speed and showing a characteristic pressure to time profile. Once transmitted to the tissue, the shockwave is translated into biochemical signals by a mechanism known as mechanotransduction. Recent experimental work exhibited a variety of cellular and molecular systems involved in the working mechanism of shockwaves. It was shown that shockwaves enhance cell proliferation via ATP-triggered extracellular signal-regulated kinase (ERK) activation resulting in improved wound healing, potentially preserving the multipotency of stem cells. Additionally growth factors and cytokines are upregulated in response to ESWT, leading among others to angiogenesis, modulation of the inflammatory process, and recruitment of endogenous stem cells.

- **Low-Intensity Pulsed Ultrasound (LIPUS)**

LIPUS has been reported in clinical case series to support osseous healing particularly in mechanical stable fractures with a fracture gap less than 1 cm. It is proposed that various stages of the healing cascade are influenced by ultrasound which includes the differentiation of mesenchymal stem cells into chrondrocytes and osteoblasts. Based on the noninvasive character of the therapy, undesired side effects may be minimal making this method suitable for ambulatory care. However, daily application (once per day for 20 min) for a period of 60 to 120 days requires a certain amount of patience and compliance of the patient. Currently, no prospective randomized controlled trials are available, making the clinical evidence of LIPUS uncertain.

- **Pulsed Electromagnetic Stimulation (PEMS)**

PEMS has been reported to affect different aspects of bone repair. These include increased blood supply, stimulated calcification of the fibrocartilage in the fracture gap, and inhibitory properties during the resorptive phase. Clinically applied, PEMS has been suggested to reach success rates of up to 80% in the treatment of nonunions. However, similar to LIPUS treatment, it is time consuming requiring high compliance of the patient.

> **Take-Home Message**
>
> Delayed fracture healing as well as nonunions is rare but still very challenging for both the patient and the surgeon. A precise evaluation of the underlying pathology and direct and indirect factors associated with impaired bony healing is of paramount importance. Based on this, the appropriate therapeutic approach should be meticulously planned actively including (involving) the patient into the discussion. The therapeutic regimen in most cases will be multimodal, involving different means such as exchange of the hardware, autologous bone grafting, and biophysical stimulation of the fracture site. Additionally, further research is required to improve methods and to develop new or optimized strategies for nonunion treatment.

References

1. Brinker MR, O'Connor DP. The incidence of fractures and dislocations referred for orthopaedic services in a capitated population. J Bone Joint Surg Am. 2004;86-A(2):290–7.
2. Court-Brown CM, Caesar B. Epidemiology of adult fractures: a review. Injury. 2006 37(8):691–7.
3. Pearce AI, Richards RG, Milz S, Schneider E, Pearce SG. Animal models for implant biomaterial research in bone: a review. Eur Cell Mater. 2007;13:1–10.
4. Aerssens J, Boonen S, Lowet G, Dequeker J. Interspecies differences in bone composition, density, and quality: potential implications for in vivo bone research. Endocrinology. 1998;139(2):663–70.
5. O'Loughlin PF, Morr S, Bogunovic L, Kim AD, Park B, Lane JM. Selection and development of preclinical models in fracture-healing research. J Bone Joint Surg Am. 2008;90(Suppl 1):79–84.
6. Martini L, Fini M, Giavaresi G, Giardino R. Sheep model in orthopedic research: a literature review. Comp Med. 2001;51(4):292–9.
7. Turner CH, Roeder RK, Wieczorek A, Foroud T, Liu G, Peacock M. Variability in skeletal mass, structure, and biomechanical properties among inbred strains of rats. J Bone Miner Res. 2001;16(8):1532–9.
8. Wang X, Mabrey JD, Agrawal CM. An interspecies comparison of bone fracture properties. Biomed Mater Eng. 1998;8(1):1–9.
9. Nunamaker DM. Experimental models of fracture repair. Clin Orthop Relat Res. 1998;(355 Suppl):S56–65.

10. Park SH, O'Connor K, Sung R, McKellop H, Sarmiento A. Comparison of healing process in open osteotomy model and closed fracture model. J Orthop Trauma. 1999;13(2):114–20.
11. Bonnarens F, Einhorn TA. Production of a standard closed fracture in laboratory animal bone. J Orthop Res. 1984;2(1):97–101.
12. Newton CD, Nunamaker DM. Textbook of small animal orthopaedics. JB Lippincott, Philadelphia. 1985.
13. Shapiro F. Cortical bone repair. The relationship of the lacunar-canalicular system and intercellular gap junctions to the repair process. J Bone Joint Surg Am. 1988;70(7):1067–81.
14. Hulse, D, Hyman, B. Fracture biology and biomechanics. in: D Slatter (Ed.) Textbook of small animal surgery. 2nd Ed. W.B. Saunders, Philadelphia; 1993:1595–1603.
15. Einhorn TA. The cell and molecular biology of fracture healing. Clin Orthop Relat Res. 1998;(355 Suppl):S7–21.
16. Gerstenfeld LC, Cullinane DM, Barnes GL, Graves DT, Einhorn TA. Fracture healing as a post-natal developmental process: molecular, spatial, and temporal aspects of its regulation. J Cell Biochem. 2003;88(5):873–84.
17. Sfeir C, Ho L, Doll BA, Azari K, Hollinger JO. Fracture repair. In: Lieberman JR, Friedlaender GE, editors. Bone regeneration and repair. Humana Press; Totowa, NJ: 2005;pp. 21–44.
18. Kon T, Cho TJ, Aizawa T, Yamazaki M, Nooh N, Graves D, et al. Expression of osteoprotegerin, receptor activator of NF-kappaB ligand (osteoprotegerin ligand) and related proinflammatory cytokines during fracture healing. J Bone Miner Res. 2001;16(6):1004–14.
19. Dimitriou R, Tsiridis E, Giannoudis PV. Current concepts of molecular aspects of bone healing. Injury. 2005;36(12):1392–404.
20. Cho TJ, Gerstenfeld LC, Einhorn TA. Differential temporal expression of members of the transforming growth factor beta superfamily during murine fracture healing. J Bone Miner Res. 2002;17(3):513–20.
21. Chen Y, Alman BA. Wnt pathway, an essential role in bone regeneration. J Cell Biochem. 2009;106(3):353–62.
22. Barnes GL, Kostenuik PJ, Gerstenfeld LC, Einhorn TA. Growth factor regulation of fracture repair. J Bone Miner Res. 1999;14(11):1805–15.
23. Bassett, CAL. (1971) Biophysical principles effecting bone structure, in Biochemistry and Physiology of Bone, 2nd edn, vol. III (ed. G.H. Bourne), Academic Press, New York, p. 1–76.
24. Deschaseaux F, Sensebe L, Heymann D. Mechanisms of bone repair and regeneration. Trends Mol Med. 2009;15(9):417–29.
25. Sfeir C, Ho L, Doll BA, Azari K, Hollinger JO. Fracture repair. In: Lieberman JR, Friedlaender GE, editors. Bone regeneration and repair. Humana Press; Totowa, NJ: 2005;pp. 21–44.
26. Mathew G, Hanson BP. Global burden of trauma: need for effective fracture therapies. Indian J Orthop. 2009;43(2):111–6.
27. Weber BG, Cech O. Pseudarthrosis: pathology, biomechanics, therapy, results. Hans Huber Medical Publisher, Berne, Switzerland. 1976.
28. Ren Y, Liu B, Feng Y, Shu L, Cao X, Karaplis A, et al. Endogenous PTH deficiency impairs fracture healing and impedes the fracture-healing efficacy of exogenous PTH(1-34). PLoS One. 2011;6(7):e23060.
29. Giannoudis PV, Einhorn TA, Marsh D. Fracture healing: the diamond concept. Injury. 2007;38(Suppl 4):S3–6.

Oral and Maxillofacial Aspects of Bone Research

Arno Wutzl

8.1 Development of the Facial Skeleton – 126

8.2 Periodontitis: Defense Mechanisms Versus Bacterial Invaders – 127

8.3 Medication-Related Osteonecrosis of the Jaws: The Phossy Jaw of the Twenty-First Century – 129

8.4 Bone Morphogenetic Proteins: The Bone Scientist's Alchemic Compound – 131

References – 134

What You Will Learn in This Chapter

Cleft lip and palate is the genetic malformation with the highest incidence of 1:700 of the face. It is the non-fusion of the body's natural structures that form before birth and affects either the lip, the alveolus, or the palate. Periodontitis is the disease of the supporting tissues of the teeth leading to destruction of the alveolus and secondary edentulism. Osteonecrosis of the jaws related to medication like bisphosphonates (BP) and the RANKL-RANK antagonist denosumab influences bone metabolism and may lead to jaw necrosis, infection, and fracture. Since bone morphogenetic proteins (BMPs) increase the differentiation and proliferation of the mesenchymal stem cell line of osteoblasts, recently they became therapeutic targets to promote bone regeneration. BMPs are already applied for elective surgery in bone regenerative surgical procedures. The efficacy of these surgical procedures is of great scientific concern.

8.1 Development of the Facial Skeleton

The facial skeleton protects the brain and the sense organs of smell, sight, and taste and makes the act of eating, facial expression, breathing, and speech possible. The mandible is a mobile bone of the facial skeleton, and since it houses the lower teeth, its motion is essential for mastication. It is formed by intramembranous ossification. The maxilla houses the teeth, forms the roof of the oral cavity, forms the nasal cavity, houses the maxillary sinus, and forms the floor of the orbit. Two maxillary bones are joined in the midline to form the middle third of the face. The zygoma forms the lateral portion of the orbit and the anterior zygomatic arch, from which the masseter muscle is suspended. The frontal bone houses the sinuses and forms the roof of the ethmoid sinuses, nose, and orbit. The paired nasal bones confine the nasal cavity [1].

The development of the face is coordinated by complex morphogenetic events and is highly susceptible to environmental and genetic factors, explaining the high incidence of facial malformations. During the first 6–8 weeks of pregnancy, the shape of the embryo's head is formed.

The frontonasal prominence grows from the top of the head down toward the future upper lip. The maxillary prominence grows from the cheeks, which meet the first lobe to form the upper lip. The mandibular prominence grows just below; two additional lobes grow from each side, which form the chin and lower lip. If these tissues fail to meet, a gap appears where the tissues should have fused. The resulting birth defect reflects the severity of individual fusion failures (e.g., from a small lip or palate fissure up to a completely malformed face).

The upper lip is formed earlier than the palate, from the first three lobes. Formation of the palate is the last step in joining the five embryonic facial lobes and involves the back portions of the maxillary and mandibular lobes. These back portions are called palatal shelves, which grow toward each other until they fuse in the middle [2]. This process is very vulnerable to multiple toxic substances, environmental pollutants, and nutritional imbalance. The biologic mechanisms of mutual recognition of the two cabinets, and the way they are glued together, are quite complex and obscure despite intensive scientific research [3].

Cleft lip and cleft palate, which can also occur together as cleft lip and palate, are variations of a type of cleft malformation caused by abnormal facial development during gestation. A cleft is a fissure or opening—a gap. Approximately 1 in 700 children born

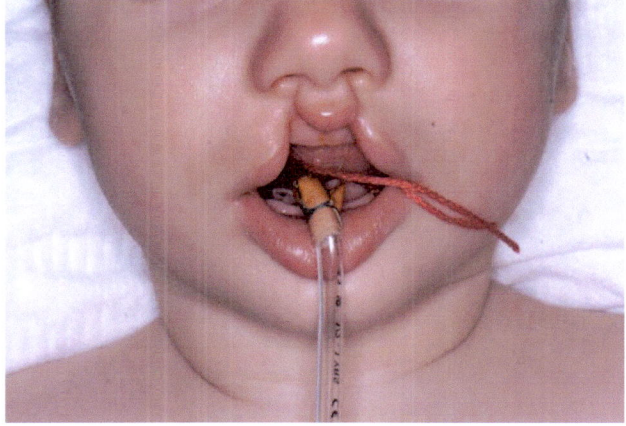

Fig. 8.1 Newborn with cleft lip and palate intubated for surgical closure of the lip

have a cleft lip or a cleft palate or both [4]. Treatment includes plastic and reconstructive surgery with rehabilitation of the bony and soft tissue structures of the alveolus and the face (Fig. 8.1).

Regardless whether bones are typically long bones, e.g., the mandible, or are of rather flat appearance such as skull bones, their external part is built as a compact structure (cortical bone), whereas their interior is a three-dimensional network of fine trabeculae (cancellous bone) [5]. Approximately 80% of the adult human skeleton consists of cortical bone and 20% of trabecular bone. Trabecular bone is considered metabolically far more active than cortical bone. Histologically, woven and lamellar bone can be distinguished; woven bone is typically formed during fetal and postnatal skeletal development. In the adult organism, only lamellar bone is formed, except under some conditions of enhanced bone formation such as fracture healing, where predominantly woven bone is built [6].

The adult skeleton bone is constantly remodeled throughout life. Bone regeneration results from bone absorption and formation and therefore depends on the generation of osteoclasts and osteoblasts. An orderly supply of these cells is essential, as changes in their activity or numbers affect bone homeostasis, causing impairment of fracture healing, periodontitis, and alveolar atrophy of the jaw.

In the following sections, selected current scientific topics of bone biology relevant to oral and maxillofacial surgery will be discussed.

8.2 Periodontitis: Defense Mechanisms Versus Bacterial Invaders

Periodontitis is defined as an inflammatory disease of the supporting tissues of the teeth caused by specific microorganisms, resulting in progressive destruction of the periodontal ligament and alveolar bone with pocket formation, recession, or both. The clinical feature of periodontitis is the presence of clinically detectable alveolar bone loss. Periodontitis is regarded as the second most common dental disease worldwide (after dental decay). In the United States, periodontitis has a prevalence of 30–50% of the population; nevertheless, only about 10% have severe forms. Chronic periodontitis affects about 750 million people or about 10.8% of the population as of 2010 [7].

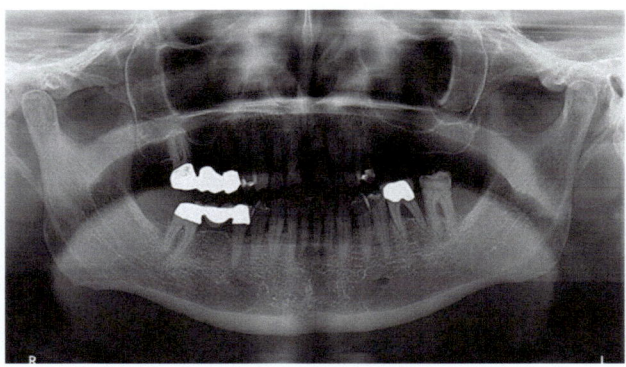

Fig. 8.2 Patient with clinical signs of periodontitis: bone destruction visible in the x-ray

Signs include redness or bleeding of gums and halitosis, or bad breath. Symptoms are gingival recession, deep pockets between the teeth and the gums, and loose teeth (in the later stages). Patients should realize that gingival inflammation and bone destruction are largely painless (Fig. 8.2) [4].

Periodontitis can be classified in early onset, adult onset, and necrotizing forms. In addition, periodontitis may be associated with systemic conditions such as diabetes and HIV infection. Sometimes it is accompanied with endodontic lesions or can be acquired with predisposing teeth anatomic factors and dental restorations [8].

The identification of bacterial pathogens in periodontal diseases has given a severe impact in the understanding of periodontitis [9]. The periodontal microbiota is a complex community of microorganisms and some of them function as pathogens.

Dental plaque is a host-associated biofilm that has been recognized to alter the properties of the microorganisms. Although the biofilm and its containing microorganisms in one patient may produce only gingivitis and in the other periodontitis, the literature shows an influence of these compounds on alveolar destruction. The inflammatory responses in the periodontal tissues induced by microorganisms in dental plaque are modulated by the hosts' immune system. The answer is an activation of the immune system including cells and complement. Transendothelial migration is of equal importance as other leukocyte functions. Specific responses include clonal selection and expansion. $CD8^+$ and $CD4^+$ positive cells contribute as well as B-cells and antibodies. Patients with genetic disorders clearly exhibit a lack of sufficient immune response leading to progressive alveolar destruction due to periodontitis [8].

Periodontitis has been linked to increased inflammation in the body, as indicated by raised levels of C-reactive protein and interleukin-6 [10]. Periodontitis thus is linked to an increased risk of stroke [11], myocardial infarction [12], and atherosclerosis [13]. Individuals with impaired fasting glucose and diabetes mellitus have a higher degree of periodontal inflammation and often have difficulties with balancing their blood glucose level owing to the constant systemic inflammatory state, caused by the periodontal inflammation [14]. Although no causative connection was proved yet, a recent study revealed an epidemiological association between chronic periodontitis and erectile dysfunction [15].

The cornerstone of successful periodontal treatment starts with establishing excellent oral hygiene, brushing with daily flossing. Persons with periodontitis must realize that it is a chronic inflammatory disease and a lifelong regimen of excellent hygiene and professional maintenance care with a dentist/hygienist or periodontist is required to maintain affected teeth. If the teeth are already lost, surgical procedures to restore the masticatory system are necessary. Part of the treatment is the augmentation and replacement of teeth with dental implants.

8.3 Medication-Related Osteonecrosis of the Jaws: The Phossy Jaw of the Twenty-First Century

In 2003 oral and maxillofacial surgeons were the first clinicians to recognize and report cases of non-healing, exposed necrotic bone in the maxillofacial region in patients treated with intravenous bisphosphonates [16–18]. Especially patients with malignant disease and metastatic cancer, but also osteoporosis patients therefore, are highly susceptible for bisphosphonate-related osteonecrosis of the jaw (BRONJ) (◘ Figs. 8.3, 8.4, and 8.5).

Ruggiero et al. [19] reported osteonecrosis of the jaws associated with bisphosphonate treatment in patients with malignant diseases or osteoporosis. Hellstein et al. [20] compared the «bis-phossy jaw» with the historical disease «phossy jaw» found in workers exposed to white phosphorus vapor during the nineteenth century. Marx et al. [21] suspected that the mechanism of osteonecrosis includes bisphosphonate-related apoptosis of osteoclasts as well as its antiangiogenic effect. The rich vascularization of the jaws renders them highly susceptible to the deposition of bisphosphonates. If bone remodeling of the jaws is depressed, this may lead to a spontaneous breakdown or to poor healing after tooth extraction.

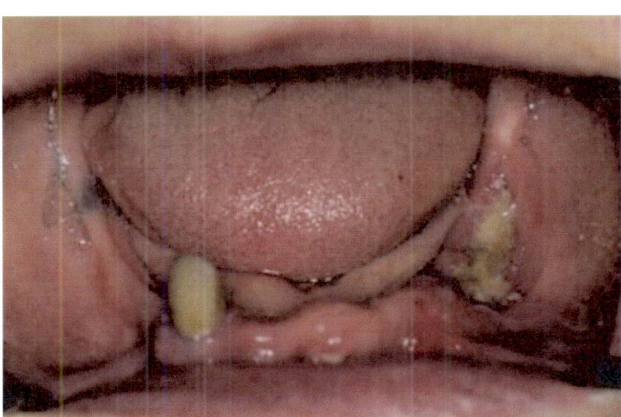

◘ Fig. 8.3   BRONJ stage 1

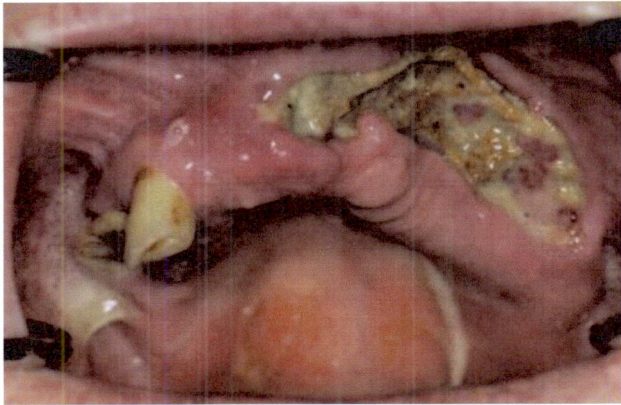

◘ Fig. 8.4   BRONJ stage 2

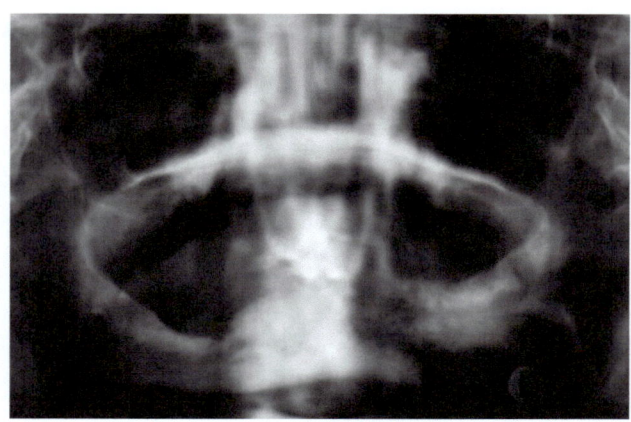

Fig. 8.5 BRONJ stage 3

Bisphosphonates have a brief half-life of 30 min to 2 h in blood. 20–80% of the substance is deposited in the bone [22]. Although the bioavailability of bisphosphonates is much lower when administered orally, the risk of BRONJ still exists in long-term therapy [23]. The prevalence of BRONJ in this setting ranges between 0.01 and 0.06% [24]. The incidence of osteonecrosis of the jaws under pamidronate or zoledronic acid was found to increase from ≤1.5% after 12 months of treatment to 7.7% after interval therapy for 3–4 years [25]. Oral trauma and poor prosthetic care are risk factors for osteonecrosis [26]. In 2007 a consensus paper was published [27] and strategies for the prevention and treatment of BRONJ determined. Higher rates are observed in patients who have undergone dental extraction or have untreated periodontitis in conjunction with poor oral hygiene [28]. In general, treatment strategy is the limitation of the affected bone with antibiotics and mouth rinsings. Surgery is performed for sequestrectomy and primary wound closure is the aim of treatment [29].

Starck and Epker (1995) described a patient who had been treated with five interforaminal implants and was given etidronate to treat osteoporosis. A peri-implant lesion developed after 5 months of treatment with oral bisphosphonates and eventually led to loss of the dental implants [30]. Bisphosphonate-related osteonecrosis of the jaw (BRONJ) after the insertion of dental implants has been reported increasingly often from 2004 onward [31]. Given the fact that the number of dental implants is dramatically increasing worldwide, guidelines of the American Association of Oral and Maxillofacial Surgeons (AAOMS) were published: dental implants should not be inserted during intravenous treatment with bisphosphonates (BP) in malignant diseases. Dental implants may be safely inserted in patients undergoing treatment with oral bisphosphonates for less than 3 years [32]. Despite initially successful osseointegration, bisphosphonate therapy may damage the dental implants several years later. Thus, peri-implantitis may precipitate the occurrence of osteonecrosis under the impact of bisphosphonates. Adherence to optimal oral hygiene is essential for the prevention of osteonecrosis [33].

Despite some negative reports about the insertion of dental implants under bisphosphonate therapy, bisphosphonates have been found to exert favorable effects on the implantation of hip joint prostheses [34]. On the other hand, an in vitro study has demonstrated a beneficial effect of bisphosphonates on osteoblast proliferation and maturation

[35]. Given a likely ONJ incidence of 0.03–0.05% during oral administration of bisphosphonates [36], one may justifiably conclude that implantation is permissible during oral administration.

Marx et al. recommend a preoperative drug holiday [37] before operations are performed in BRONJ patients. However, recent investigations have shown that the interruption of bisphosphonate therapy has no impact on the risk profile [38]. When surgery is performed at a time of high bisphosphonate levels, one may anticipate a poor outcome.

According to Marx et al., serum CTx (C-terminal telopeptide) levels may be used to measure the course of bisphosphonate therapy [37]. CTx is a specific marker of type-I collagen. CTx is released by the organic matrix during bone resorption. The test is directly proportional to the resorptive activity of osteoclasts and permits measurement of BRONJ remodeling rates. The CTx levels reflect the risk of osteonecrosis because the long half-life of bisphosphonates in BRONJ and their prolonged aftereffects must be taken into account [38].

Recently an association of osteonecrosis of the jaw not only with bisphosphonates but also with other drugs has been reported; the term «drug-induced osteonecrosis» of the jaws (DIONJ) was coined [39]. A new therapeutic path to affect pathologic bone turnover is to block the receptor activator of nuclear factor-κB ligand (RANKL) with the monoclonal IgG2 antibody denosumab. Pharmacologically denosumab mimics the effect of osteoprotegerin on RANKL and rapidly leads to a reduction in bone turnover and an increase in bone density. Until now, denosumab has been well tolerated in all clinical trials, but reports of denosumab-related osteonecrosis of the jaws in patients with cancer underline recently given concerns about this drug. Interestingly, also an association of osteonecrosis of the jaws with angiogenesis inhibitors (such as bevacizumab) has been discussed [40].

Bone morphogenetic proteins are exceptional compounds with a medical importance for the facial skeleton.

8.4 Bone Morphogenetic Proteins: The Bone Scientist's Alchemic Compound

Recently, bone morphogenetic proteins (BMPs) became interesting therapeutic targets to promote bone regeneration. They are members of the TGF-β super family of polypeptides, which have a conserved carboxyterminal region containing seven cysteine repeats [41]. More than 40 BMP-homologous molecules have been described so far [42]. BMPs are capable of inducing ectopic bone formation in vivo when implanted into muscular sites [43]. Accumulating evidence indicates that the BMP family plays an important role in organogenesis of the lung, heart, teeth, gut, skin, and kidney as well as the bone and cartilage [44]. In vitro studies showed that BMPs exert a wide range of biological responses such as regulation of cell proliferation, differentiation, and apoptosis in a variety of cell types. BMPs induce the differentiation of undifferentiated mesenchymal cells into osteogenic cells and enhance the differentiated function of osteoblasts [45]. As a BMP-induced endochondral ossification process, undifferentiated mesenchymal cells first differentiate into chondrocytes that lay down a cartilaginous structure, which then is resorbed by osteoclasts and replaced with bone by osteoblasts [46]. Based on these sequential biologi-

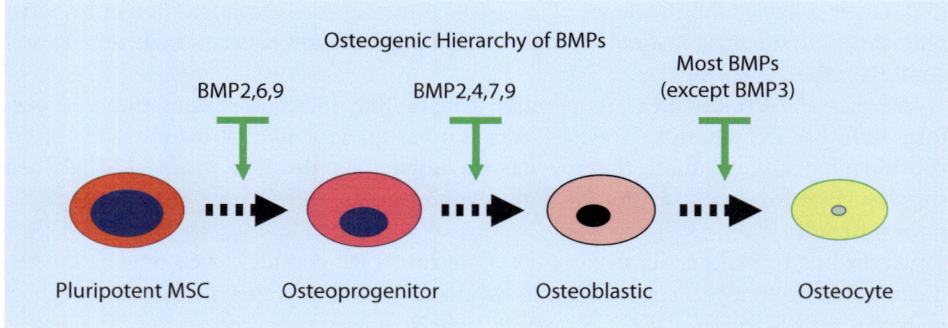

Fig. 8.6 An in vitro study by Cheng et al. [52] evaluated the osteogenic activity of fourteen types of BMPs on osteoblastic progenitor cells with an osteogenic hierarchical model in which BMP-2 and BMP-6 may play an important role in inducing osteoblast differentiation

cal events, it may be speculated that BMPs may act on osteoclastic bone resorption as well as cartilage and bone formation, because bone resorption is an essential process in bone development, maintenance, and regeneration [47].

Recently, several recombinant forms of BMPs, especially BMP-2 and BMP-7, have been shown to induce bone formation in vivo and both have been tested in clinical trials [48–51]. Data from these trials suggested that BMP-induced bone formation is at least equivalent to autogenous bone grafting when used in alveolar bone grafting, spinal fusions, and tibial nonunion. A number of studies have confirmed the effectiveness of BMPs in promoting osteogenesis (◘ Fig. 8.6).

BMPs have shown varying degrees of success in the clinical setting, and further studies on their mechanisms of action and optimal formulations are required to optimize their osteopromotive effectiveness.

For example, the in vivo effect of BMP-5 and its role in bone formation have been recently investigated. As the growth of long bones and certain stages of fracture healing occur by endochondral bone formation, these studies suggest that BMP-5 plays an important role in regulating the cellular activity involved in the bone remodeling process [53]. Its effects on bone cells in vitro have not been reported thus far, but our data indicate the significant role of BMP-5 in the bone remodeling process as well [54]. BMP-6-deficient mice are largely unremarkable, with the exception of a defect of the sternum [55]. Hence, the osteogenic potency of each BMP might depend on the cell line, the stage of differentiation of cells, and the concentration of each BMP [56]. The exact combination and dosage of BMPs used in a clinical setting might improve the differentiation of these mesenchymal cells and thus promote fracture healing as well as osteogenic procedures. BMPs are already in clinical use. The first indication in that BMPs were used was the nonunion of tibial fractures [50]. Also of great advantage was the use of BMPs for the treatment of spinal fusions secondary to osteoporotic fractures of the spine [57]. Since BMPs proved to be a safe treatment method in orthopedic indications, in oral and maxillofacial surgery, dental implants were coated with BMPs for faster osseointegration [58] and bone augmentation procedures [59]. These augmentation procedures may be performed to place implants or after ablative surgery to prevent microvascular free flap procedures [60].

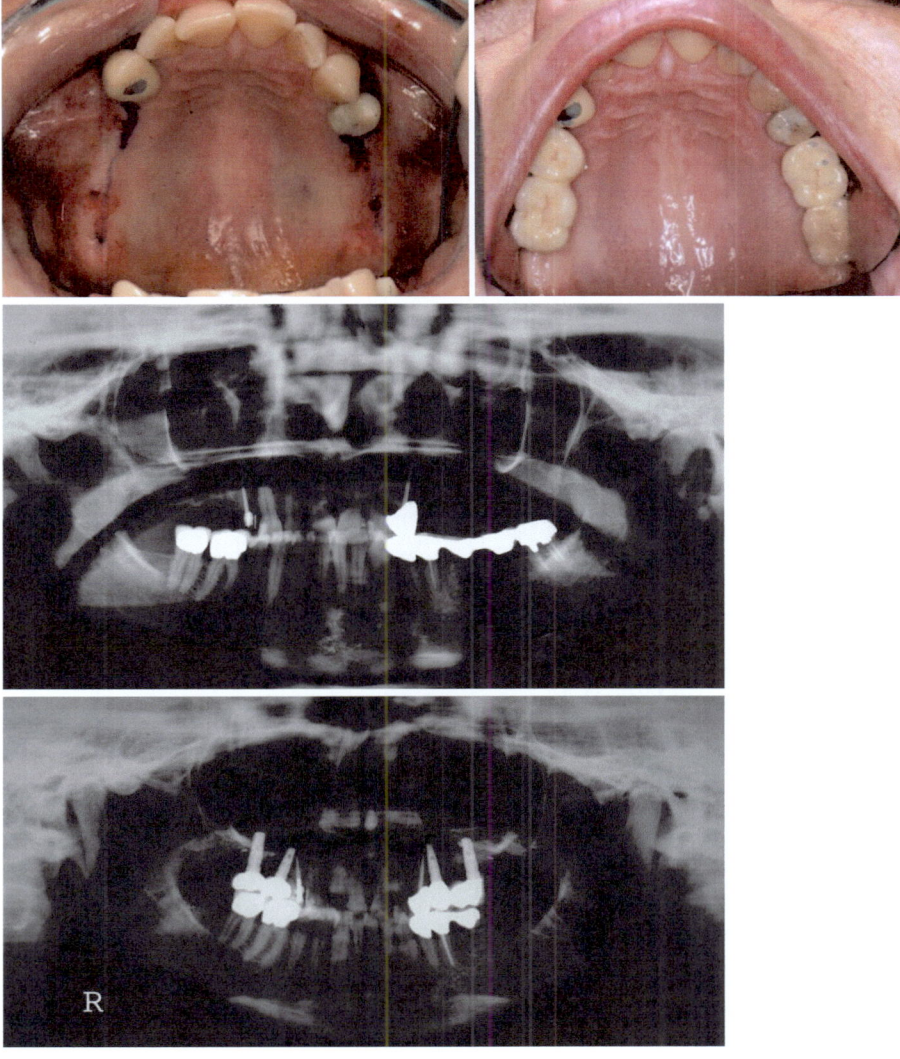

Fig. 8.7 Patient before and after a sinus lift operation followed by the insertion of dental implants and loading with two artificial molars on both sides in the upper jaw

Currently BMP-2 and BMP-7 may be used in vivo. Clinical investigations concern the mode of application (either with collagen membranes or with a virus used as a vector), the use with autogenous bone transfer to activate local stem cells, and the dosage [61].

Just recently, BMPs were also applied in elective surgery of the face, and BMPs were used to improve sinus lift operations for the later application of dental implants [62] (Fig. 8.7).

> **Take-Home Message**
> - Disturbances of the facial skeleton include genetic malformations like cleft lip and palate.
> - Bacterial invasion versus osteoimmunologic response is seen in periodontitis. Bisphosphonates, denosumab, and related medication, meant to reverse bone aging, may lead to osteonecrosis of the jaws.
> - The application of regenerative compounds like the BMPs may lead to the reversal of bone malformations, aging, and destruction. BMPs promise an improvement in reconstructive surgery of the bone augmentation techniques of the face.
> - The facial skeleton protects the brain; houses the human senses of smell, sight, and taste; and gives us the opportunity to hear. It gives form to the face and determines our appearance. Disturbances have a great influence on our ability of communication with our environment and getting in contact with our beloved. Therefore mankind expends great scientific strength to maintain and improve its beauty and grace.

References

1. Medscape.
2. Hillig U. Lippen-Kiefer-Gaumen-Spalten: Klassifikation und Epidemiologie. Springer Verlag Fortschritte der Kieferorthopädie. 1991;52(4):230–6.
3. Schwenzer N, Ehrenfeld M, Zahn- Mund-, Kiefer- und Heilkunde, Mund- Kiefer- Gesichtschirurgie, 4. vollständig überarbeitete und erweiterte Auflage, Kapitel 6 Lippen- Kiefer- Gaumenspalten, Georg Thieme Verlag KG. Stuttgart/New York. 1981, 2011.
4. Wikipedia.
5. Baron R. Anatomy and ultrastructure of bone. In: Favus MJ, editor. Primer on the metabolic bone diseases and disorders of mineral metabolism, vol. 3. 4th ed. Philadelphia: Lippincott Williams & Wilkins; 1999.
6. Fleisch H. Bisphosphonates in bone disease. From the laboratory to the patient. 3rd ed. New York: Parthenon Publishing Group; 1997.
7. Vos T, Flaxman AD, Naghavi M, Lozano R, Michaud C, Ezzati M, Shibuya K, Salomon JA, Abdalla S, Aboyans V, Abraham J, Ackerman I, Aggarwal R, Ahn SY, Ali MK, Almazroa MA, Alvarado M, Anderson HR, Anderson LM, Andrews KG, Atkinson C, Baddour LM, Bahalim AN, Barker-Collo S, Barrero LH, Bartels DH, Basáñez M-G, Baxter A, Bell ML, et al. Years lived with disability (YLDs) for 1160 sequelae of 289 diseases and injuries 1990-2010: a systematic analysis for the global burden of disease study 2010. Lancet. 2012;380(9859):2163–96.
8. Newman GN, Takei HH, Carranza FA. Clinical Periodontology. PA, Philadelphia: WB Saunders Company; 2002.
9. Socransky SS, Haffajee AD. The bacterial etiology of destructive periodontal disease: current concepts. J Periodontol. 1992;63:322.
10. D'Aiuto F, Ready D, Tonetti MS. Periodontal disease and C-reactive protein-associated cardiovascular risk. J Periodontal Res. 2004;39(4):236–41.
11. Pussinen PJ, Alfthan G, Jousilahti P, Paju S, Tuomilehto J. Systemic exposure to Porphyromonas gingivalis predicts incident stroke. Atherosclerosis. 2007;193(1):222–8.
12. Pussinen PJ, Alfthan G, Tuomilehto J, Asikainen S, Jousilahti P. High serum antibody levels to Porphyromonas gingivalis predict myocardial infarction. Eur J Cardiovasc Prev Rehabil. 2004;11(5):408–11.
13. Ford PJ, Gemmell E, Timms P, Chan A, Preston FM, Seymour GJ. Anti-P. Gingivalis response correlates with atherosclerosis. J Dent Res. 2007;86(1):35–40.
14. Zadik Y, Bechor R, Galor S, Levin L. Periodontal disease might be associated even with impaired fasting glucose. Br Dent J. 2010;208(10):E20.

15. Zadik Y, Bechor R, Galor S, Justo D, Heruti RJ. Erectile dysfunction might be associated with chronic periodontal disease: two ends of the cardiovascular spectrum. J Sex Med. April 2009;6(4):1111–6.
16. Marx RE. Pamidronate (Aredia) and zoledronate (Zometa) induced avascular necrosis of the jaws: a growing epidemic [letter]. J Oral Maxilofac Surg. 2003;61:1115–8.
17. Migliorati CA. Bisphosphanates and oral cavity cancer avascular bone necrosis (correspondence). J Clin Oncol. 2003;21:4253–4.
18. Wang J, Goodger NM, Pogrel MA. Osteonecrosis of the jaws associated with cancer chemotherapy. J Oral Maxillofac Surg. 2003;6:1104–7.
19. Ruggiero SL, Mehrotra B, Rosenberg TJ, Engroff S. Osteonecrosis of the jaws associated with the use of bisphosphonates: a review of 63 cases J Oral Maxillofac Surg. 2004;62:527–34.
20. Hellstein JW, Marek CL. Bisphosphonate osteochemonecrosis (bis-phossy jaw): is this phossy jaw of the 21st century? J Oral Maxillofac Surg. 2005;63:682–9.
21. Khan AA, Morrison A, Hanley DA, Felsenberg D, McCauley LK, O'Ryan F, Reid IR, Ruggiero SL, Taguchi A, Tetradis S, Watts NB, Brandi ML, Peters E, Guise T, Eastell R, Cheung AM, Morin SN, Masri B, Cooper C, Morgan SL, Obermayer-Pietsch B, Langdahl BL, Al Dabagh R, Davison KS, Kendler DL, Sándor GK, Josse RG, Bhandari M, El Rabbany M, Pierroz DD, Sulimani R, Saunders DP, Brown JP, Compston J, International Task Force on Osteonecrosis of the Jaw. Diagnosis and management of osteonecrosis of the jaw: a systematic review and international consensus. J Bone Miner Res. 2015;30(1):3–23. doi:10.1002/jbmr.2405. Review.
22. Fleisch H. Introduction to bisphosphonates. History and functional mechanisms. Orthopade. 2007;36(2):103–4. 106-9
23. Lo JC, et al. Prevalence of osteonecrosis of the jaw in patients with oral bisphosphonate exposure. Journal of oral and maxillofacial surgery : officia journal of the American Association of Oral and Maxillofacial Surgeons. 2010;68(2):243–53.
24. Ruggiero SL, et al. American Association of Oral and Maxillofacial Surgeons position paper on bisphosphonate-related osteonecrosis of the jaw - 2009 update. Aust Endod J. 2009;35(3):119–30.
25. Bamias A, Kastritis E, Bamia C, Moulopoulos L, Melakopoulos I, Bozas G, et al. Osteonecrosis of the jaw in cancer after treatment with bisphosphonates: incidence and risk factors. J Clin Oncol. 2005;23: 8580–7.
26. Mücke T, Krestan CR, Mitchell DA, Kirschke JS, Wutzl A. Bisphosphonate and medication-related osteonecrosis of the jaw: a review. Semin Musculoskelet Radiol. 2016 Jul;20(3):305–14.
27. Advisory Task Force on Bisphosphonate-Related Ostenonecrosis of the Jaws, American Association of Oral and Maxillofacial Surgeons. American Association of Oral and Maxillofacial Surgeons position paper on bisphosphonate-related osteonecrosis of the jaws. J Oral Maxillofac Surg. 2007;65:3 69–76.
28. Svejda B, Muschitz C, Gruber R, Brandtner C, Svejda C, Gasser RW, Santler G, Dimai HP. Positionspapier zur medikamentenassoziierten Osteonekrose des Kiefers (MRONJ). Wien Med Wochenschr. 2016;166:68–74.
29. Holzinger D, Seemann R, Klug C, Ewers R, Millesi G, Baumann A, Wutzl A. Long-term success of surgery in bisphosphonate-related osteonecrosis of the jaws (BRONJs). Oral Oncol. 2013 Jan;49(1):66–70.
30. Starck WJ, Epker BN. Failure of osseointegrated dental implants after bisphosphonate therapy for osteoporosis: a case report. Int J Oral Maxillofac Implants. 1995;10(1):74–8.
31. Troeltzsch M, Cagna D, Stähler P, Probst F, Kaeppler G, Troeltzsch M, Ehrenfeld M, Otto S. Clinical features of peri-implant medication-related osteonecrosis of the jaw: Is there an association to peri-implantitis?. J Craniomaxillofac Surg. 2016. S1010–5182(16)30232–3.
32. Advisory Task Force on Bisphosphonate-Related Ostenonecrosis of the Jaws, American Association of Oral and Maxillofacial Surgeons. American Association of Oral and Maxillofacial Surgeons Position Paper on bisphosphonate-related osteonecrosis of the jaws. J Oral Maxillofac Surg. 2007;65(3):369–76.
33. Holzinger D, Seemann R, Matoni N, Ewers R, Millesi W, Wutzl A. Effect of dental implants on bisphosphonate-related osteonecrosis of the jaws. J Oral Maxillofac Surg. 2014;72(10):1937.
34. Prieto-Alhambra D, et al. Association between bisphosphonate use and implant survival after primary total arthroplasty of the knee or hip: population based retrospective cohort study. BMJ. 2011;343:d7222.
35. Im GI, et al. Osteoblast proliferation and maturation by bisphosphonates. Biomaterials. 2004;25(18):4105–15.
36. Assael LA. Bisphosphonates and oral health: primer and an update for the practicing surgeon. Oral Maxillofac Surg Clin North Am. 2011;23(3):443–53

37. Marx RE, Cillo JE Jr, Ulloa JJ. Oral bisphosphonate-induced osteonecrosis: risk factors, prediction of risk using serum CTX testing, prevention, and treatment. J Oral Maxillofac Surg. 2007;65(12):2397–410.
38. Khosla S, et al. Bisphosphonate-associated osteonecrosis of the jaw: report of a task force of the American Society for Bone and Mineral Research. J Bone Miner Res. 2007;22(10):1479–91.
39. Uyanne J, Calhoun CC, Le AD. Antiresorptive drug-related osteonecrosis of the jaw. Dent Clin N Am. 2014 Apr;58(2):369–84.
40. Sivolella S, Lumachi F, Stellini E, Favero L. Denosumab and anti-angiogenetic drug-related osteonecrosis of the jaw: an uncommon but potentially severe disease. Anticancer Res. 2013 May;33(5):1793–7.
41. Wozney JM. The bone morphogenetic protein family and osteogenesis. Mol Reprod Dev. 1992;32:160–7.
42. Yamashita H, Ten Dijke P, Heldin CH, Miyazono K. Bone morphogenetic protein receptors. Bone. 1996;19:569–74.
43. Wozney JM, Rosen V, Celeste AJ, Mitsock LM, Whitters MJ, Kriz RW, Hewick RM, Wang EA. Novel regulators of bone formation: molecular clones and activities. Science. 1988;242:1528–34.
44. Hoodless PA, Haerry T, Abdollah S, Stapleton M, O'Connor MB, Attisano L, Wrana JL. MADR1, a MAD-related protein that functions in BMP2 signaling pathways. Cell. 1996;85:489–500.
45. Yamaguchi A, Ishizuya T, Kintou N, Wada Y, Katagiri T, Wozney JM, Rosen V, Yoshiki S. Effects of BMP-2, BMP-4, and BMP-6 on osteoblastic differentiation of bone marrow-derived stromal cell lines, ST2 and MC3T3-G2/PA6. Biochem Biophys Res Commun. 1996;220:366–71.
46. Hunziker EB. Mechanism of longitudinal bone growth and its regulation by growth plate chondrocytes. Microsc Res Tech. 1994;28:505–19.
47. Baron RE. Anatomy and ultrastructure of bone. In: Favus MJ, editor. Primer on the metabolic bone diseases and disorders of mineral metabolism. 3rd ed. New York: Lippincot-Raven; 1996. p. 3–10.
48. Fujioka-Kobayashi M, Sawada K, Kobayashi E, Schaller B, Zhang Y, Miron RJ. Osteogenic potential of rhBMP9 combined with a bovine-derived natural bone mineral scaffold compared to rhBMP2. Clin Oral Implants Res. 2017;28:381–7. doi:10.1111/clr.12804.
49. Jung RE, Glauser R, Scharer P, Hammerle CH, Sailer HF, Weber FE. Effect of rhBMP-2 on guided bone regeneration in humans. Clin Oral Implants Res. 2003;14:556–68.
50. Govender S, Csimma C, Genant HK, Valentin-Opran A, Amit Y, Arbel R, Aro H, Atar D, Bishay M, Borner MG, Chiron P, Choong P, Cinats J, Courtenay B, Feibel R, Geulette B, Gravel C, Haas N, Raschke M, Hammacher E, van der Velde D, Hardy P, Holt M, Josten C, Ketterl RL, Lindeque B, Lob G, Mathevon H, McCoy G, Marsh D, Miller R, Munting E, Oevre S, Nordsletten L, Patel A, Pohl A, Rennie W, Reynders P, Rommens PM, Rondia J, Rossouw WC, Daneel PJ, Ruff S, Ruter A, Santavirta S, Schildhauer TA, Gekle C, Schnettler R, Segal D, Seiler H, Snowdowne RB, Stapert J, Taglang G, Verdonk R, Vogels L, Weckbach A, Wentzensen A, Wisniewski T, BMP-2 Evaluation in Surgery for Tibial Trauma (BESTT) Study Group. Recombinant human bone morphogenetic protein-2 for treatment of open tibial fractures: a prospective, controlled, randomized study of four hundred and fifty patients. J Bone Joint Surg Am. 2002;84-A:2123–34.
51. Delawi D, Jacobs W, van Susante JL, Rillardon L, Prestamburgo D, Specchia N, Gay E, Verschoor N, Garcia-Fernandez C, Guerado E, Quarles van Ufford H, Kruyt MC, Dhert WJ, Oner FC. OP-1 compared with iliac crest Autograft in instrumented posterolateral fusion: a randomized, Multicenter non-inferiority trial. J Bone Joint Surg Am. 2016;98(6):441–8. doi:10.2106/JBJS.O.00209.
52. Cheng H, Jiang W, Phillips FM, Haydon RC, Peng Y, Zhou L, Luu HH, An N, Breyer B, Vanichakarn P, Szatkowski JP, Park JY, He TC. Osteogenic activity of the fourteen types of human bone morphogenetic proteins (BMPs). J Bone Joint Surg Am. 2003;85-A:1544–52.
53. Lieberman JR, Daluiski A, Einhorn TA. The role of growth factors in the repair of bone. Biology and clinical applications. J Bone Joint Surg Am. 2002;84-A:1032–44.
54. Wutzl A, Brozek W, Lernbass I, Rauner M, Hofbauer G, Schopper C, Watzinger F, Peterlik M, Pietschmann P. Bone morphogenetic proteins 5 and 6 stimulate osteoclast generation. J Biomed Mater Res A. 2006;77(1):75–83.
55. Yeh LC, Adamo ML, Olson MS, Lee JC. Osteogenic protein-1 and insulin-like growth factor I synergistically stimulate rat osteoblastic cell differentiation and proliferation. Endocrinology. 1997;138:4181–90.
56. Wutzl A, Rauner M, Seemann R, Millesi W, Krepler P, Pietschmann P, Ewers R. Bone morphogenetic proteins 2, 5, and 6 in combination stimulate osteoblasts but not osteoclasts in vitro. J Orthop Res. 2010 Nov;28(11):1431–9.

57. Mummaneni PV, Haid RW, Rodts GE. Lumbar interbody fusion: state-of-the-art technical advances. Invited submission from the joint section meeting on disorders of the spine and peripheral nerves. J Neurosurg Spine. 2004;1:24–30.
58. Becker J, Kirsch A, Schwarz F, Chatzinikolaidou M, Rothamel D, Lekovic V, Laub M, Jennissen HP. Bone apposition to titanium implants biocoated with recombinant human bone morphogenetic protein-2 (rhBMP-2). A pilot study in dogs. Clin Oral Investig 2006;10(3):217–24.
59. Jensen OT, Lehman H, Ringeman JL, Casap N. Fabrication of printed titanium shells for containment of BMP-2 composite graft materials for alveolar bone reconstruction. Int J Oral Maxillofac Implants. 2014 Jan-Feb;29(1):e103–5.
60. Warnke PH, Springer IN, Wiltfang J, Acil Y, Eufinger H, Wehmöller M, Russo PA, Bolte H, Sherry E, Behrens E, Terheyden H. Growth and transplantation of a custom vascularised bone graft in a man. Lancet. 2004;364(9436):766–70.
61. Carreira AC, Lojudice FH, Halcsik E, Navarro RD, Sogayar MC, Granjeiro JM. Bone morphogenetic proteins: facts, challenges, and future perspectives. J Dent Res. 2014;93(4):335–45.
62. Kim HJ, Chung JH, Shin SY, Shin SI, Kye SB, Kim NK, Kwon TG, Paeng JY, Kim JW, Oh OH, Kook MS, Yang HJ, Hwang SJ. Efficacy of rhBMP-2/hydroxyapatite on sinus floor augmentation: a Multicenter, randomized controlled clinical trial. J Dent Res. 2015;94(9 Suppl):158S–65S. doi:10.1177/0022034515594573.

Cartilage Biology: Essentials and Methods

Sylvia Nürnberger and Stefan Marlovits

9.1 Introduction – 140

9.2 Articular Cartilage Function – 141

9.3 Articular Cartilage Morphology – 141

9.4 Subdivision of Articular Cartilage in Four Layers – 143

9.5 Compartmentalisation of the ECM in Relation to the Chondrocytes – 147

9.6 The Chondrocyte and its Pericellular Microenvironment – 148

9.7 Articular Cartilage Structure Under Load – 149

9.8 Morphological Methods for Cartilage Research – 150

Suggested Reading – 150

© Springer International Publishing AG 2017
P. Pietschmann (ed.), *Principles of Bone and Joint Research*, Learning Materials in Biosciences,
DOI 10.1007/978-3-319-58955-8_9

What You Will Learn in This Chapter

Articular cartilage is the clinically most relevant hyaline cartilage since frequently damaged by trauma or abuse but without any intrinsic potential to regenerate. It covers the ends of the bones and allows frictionless movement due to its stiff and elastic properties. The biomechanics and consecutive function of articular cartilage are determined by its composition and morphological organisation. Knowledge about the cartilage biology is essential to understand its function, potential, durability and vulnerability. This chapter explains the histological organisation and ultrastructural details of this tissue, which is dominated by extracellular matrix with a special arrangement of collagen and a high amount of proteoglycans that are produced by the chondrocytes embedded in between. The chondrocytes, even if capable to metabolise and regenerate the matrix, are not able to migrate through their dense environment. In case of tissue damage, this means that they may not approach the defect to fill gaps and regenerate degraded or lost tissue. Chondrocytes are specialised cells that persist under the hypoxic conditions of the dense matrix. Their special microenvironment and a sensory organ, the cilium, allow to detect changes in the mechanical conditions on which they react by matrix synthesis or, in contrary, by degradation. These initial changes and cartilage remodelling remain though unrecognised since the tissue is not innervated. Consequently, degradation is proceeding until reaction of the surrounding tissues in the joint cavity and finally causes pain in an advanced stage of degradation. Clinical regeneration by cell therapy and tissue engineering, which is possible in young patients, requires the development of the complex hyaline cartilage tissue which is described in this chapter.

9.1 Introduction

Cartilage is part of the skeletal system and appears mainly in relation to and connected with specific mechanical conditions. Nasal and auricular cartilages both help to shape the human body, while articular cartilage and menisci act like shock absorbers at the contact points of long bones. Depending on the tissue characteristics, three types of cartilage may be distinguished: hyaline, fibrous and elastic cartilage (◘ Fig. 9.1). They all have in common that their largest volume is formed by extracellular matrix (ECM), consisting mainly of collagen type II and proteoglycans. While fibrous cartilage has the lowest proteoglycan content in relation to collagen and elastic cartilage contains additional elastic fibres, hyaline cartilage is a dense homogenous connective tissue with no obvious fibrillation in a light microscopic overview. Hyaline cartilage is the most frequent of those three cartilage types in the human body and forms the ribs and nose and covers the ends of the long bones. Ontogenetically, hyaline cartilage has different origins; while nasal cartilage is of ectodermal origin, deriving from the neural crest, most hyaline cartilage structures such as the ribs, intervertebral discs and articular cartilage are of mesodermal origin. Articular cartilage forms at the ends of long bones from a remaining layer of the cartilaginous anlagen. It is considered to be mature when the epiphyseal plate is closed (skeletal maturity).

Hyaline cartilage is characterised by an abundance of extracellular matrix, making up more than 90% of the tissue volume and consisting mainly of collagen type II and sulphated proteoglycans. Cells make up 10% of the tissue volume and are embedded in the matrix, either individually, in pairs or in groups. They all belong to the same cell type, chondrocytes. Although few in number, chondrocytes play a crucial role in the production, organisation and maintenance of the extracellular matrix and, in turn, are protected from the high forces of mechanical load and shear by the extracellular matrix they are embedded in.

Articular cartilage is an especially clinically relevant hyaline cartilage tissue as trauma and age-related abuse require intervention and, in the best case, regeneration. Therefore, even if all hyaline tissues are structurally very similar, the following description will focus on the articular cartilage.

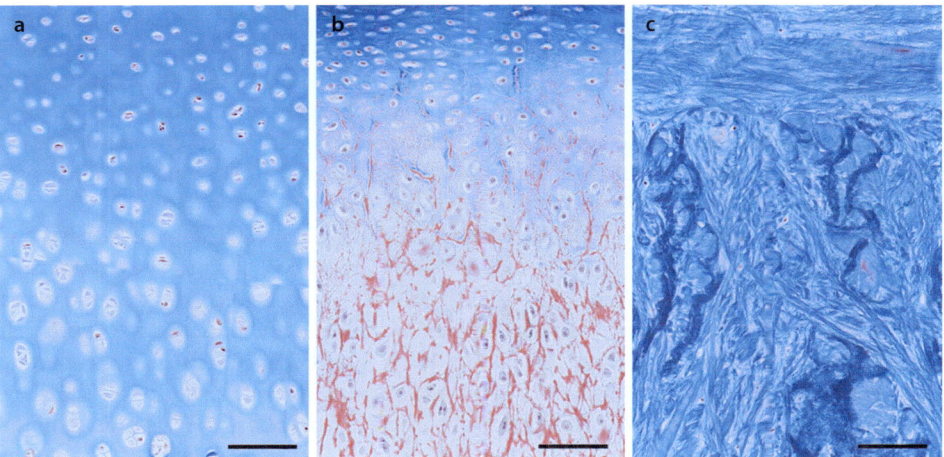

Fig. 9.1 Histological image characteristics of three different cartilage types: **a** Articular cartilage consisting of homogenous dense matrix and chondrocytes regularly distributed within. **b** In elastic cartilage, elastic fibres are embedded in the homogenous matrix forming a network around the cells. **c** Fibrous cartilage is characterized by large collagen bundles stretching through the tissue. MSB staining of paraffin sections displaying collagen in blue and elastic fibres as well as cells in red. Scale bar = 100 μm

9.2 Articular Cartilage Function

Articular cartilage is a thin but tough tissue covering the articulating surface of all diarthrodial joints. It creates a stiff but elastic interface between the articulating ends of the long bones and allows frictionless movement. Likewise it (1) protects the bone from abrasion; (2) dampens impulses and transmits high compressive loads and shearing forces to the subchondral bone; (3) creates joint congruity, reducing contact stress between apposing bones; and (4) provides a smooth, lubricated surface that facilitates movement with little friction between articulating surfaces.

These mechanobiological functions are possible due to the highly hydrated state of hyaline cartilage; interstitial water accounts for 60–80% of the wet weight. In association with proteoglycans, interstitial water is largely responsible for the load bearing and elastic properties of cartilage. Interstitial water is also a carrier system for gases, small proteins and metabolites which are exchanged by pumping the fluid through the dense matrix, driven by intermittent pressure from joint movement and loading. The dynamically pulsated interstitial water, therefore, acts as a circulatory system in this vessel-free tissue.

9.3 Articular Cartilage Morphology

In the macroscopic view, untreated articular cartilage is a homogenous, dense tissue with a whitish colour and without compartmentation or substructure. The reason for this appearance is that the tissue is dominated by the ECM which forms a tight three-dimensional network of collagen densely packed with proteoglycans. There are no blood or lymphatic vessels or nerves and only 1% of the tissue volume is formed by cells. Cellularity varies strongly between different joints, joint regions and species and increases with the body size of the animal. Cellularity further depends on age, pathological circumstances and individual differences. Variations during ageing comprise location-dependent increased or decreased cellularity. While cell death mainly affects cells in the superficial region, reducing cell numbers, cell proliferation leads to cluster formation in deeper joint regions. Despite the low cell-to-matrix ratio, chondrocytes alone are responsible for ECM synthesis and turnover. Due to their high sensitivity to mechanical stimuli, chondrocytes may adapt their ECM metabolism to the mechanical demands on the tissue.

The structural organisation of the ECM becomes visible with staining methods of histology and electron microscopy. With certain fixation methods, such as ruthenium hexammine trichloride or osmium tetroxide treatment, some aspects of the structural organisation of the tissue already become visible macroscopically. Chondrocytes and their surrounding matrix can be distinguished from the rest of the ECM. Histological staining procedures reveal a distinct organisation on two levels: (1) a subdivision of the cartilage into four layers on the basis of cell and matrix characteristics such as the orientation of the collagen fibres in relation to the joint surface (◘ Figs. 9.2, 9.3, and 9.4) and (2) a compartmentalisation of the ECM in relation to the chondrocytes (◘ Figs. 9.4 and 9.5).

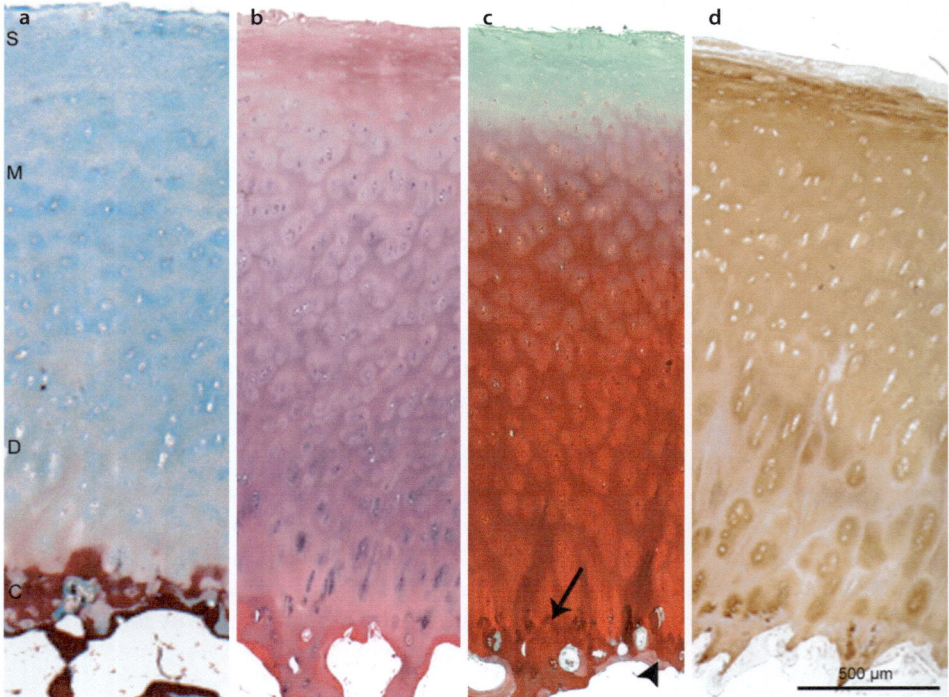

◘ **Fig. 9.2** Articular cartilage characteristics visualised by different histological stainings: **a** AZAN displays the main cartilage in *blue* and the calcified cartilage and subchondral bone in *red*. **b** Haematoxylin and eosin, one of the most common types of histological stainings, provides a general morphological overview. **c** Safranin-O staining displays proteoglycans in *red* and shows increasing staining intensity towards the bone. **d** Collagen type II immunolabelling demonstrates the nearly ubiquitous presence of collagen type II, except for the uppermost region of articular cartilage which is formed by collagen type I. *S, M, D* and *C* indicate the four layers of articular cartilage: superficial (*S*), middle (*M*), deep (*D*), and calcified (*C*) zone. *Arrow* tidemark, *arrowhead* cement line (formalin-fixed human femur condylus) (© Georg Thieme Verlag KG)

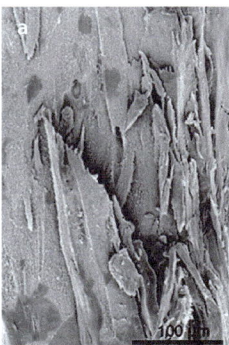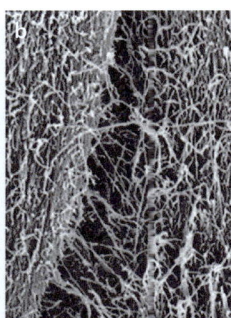

Fig. 9.3 Scanning electron microscopic image of articular cartilage after cryofracturing shows the leave-like arrangement of the collagen matrix formed from densely arranged collagen fibrils (© Georg Thieme Verlag KG)

9.4 Subdivision of Articular Cartilage in Four Layers

The structural organisation of articular cartilage may be described by differentiating four zones from the surface to the bone (Figs. 9.2, 9.4 and 9.6). These zones are characterised by collagen orientation, proteoglycan distribution and the organisation of chondrocytes at different depths throughout the matrix. However, except for the deepest layer, the calcified zone, transitions between these areas are continuous, without any sharp borders or delineations.

— *Zone I*. The superficial or tangential layer lies adjacent to the joint cavity and occupies 5–10% of the matrix volume. It is the only layer which may contain collagen type I and where collagen fibres are aligned parallel to the articular surface. Proteoglycans are rare in this layer, explaining the fibrous appearance in histology. Nevertheless, the surface is smooth and sealed by a specialised glycoprotein, lubricin. The cells are discoidal and the collagen alignment is orientated parallel to the surface. There is no particular compartmentation of the matrix around the chondrocytes in this layer, at least in humans.

— *Zone II*. The intermediate, transitional or middle layer represents 40–45% of the matrix volume. It is characterised by obliquely orientated collagen type II fibrils and a continuous increase in proteoglycan content as tissue depth increases. Chondrocytes are round or spheroidal, sometimes paired (double chondron) and surrounded by a faintly expressed pericellular matrix.

— *Zone III*. The deep or radial layer makes up 40–45% of the matrix volume. Collagen fibres (collagen type II) are arranged vertically (radially) and the interfibrillar space is densely filled with proteoglycans. In general, the density of the matrix is the highest in this region of the tissue. The chondrocytes are round and mostly arranged in columns of several cells that are surrounded by a shared, well-expressed pericellular matrix. These multichondrons are aligned along the radial collagen network and are therefore vertically arranged.

— *Zone IV*. The calcified cartilage layer forms 5–10% of the matrix volume. It is characterised by high concentrations of calcium salts (hydroxyapatite) embedded between the collagen fibres and proteoglycan. It is demarcated by the *tide mark*, separating non-calcified from calcified cartilage, and a *cement line* towards the bone. The latter is formed during endochondral ossification of the terminal growth plate at maturity. The spherical chondrocytes are arranged in stacked or spherical groups.

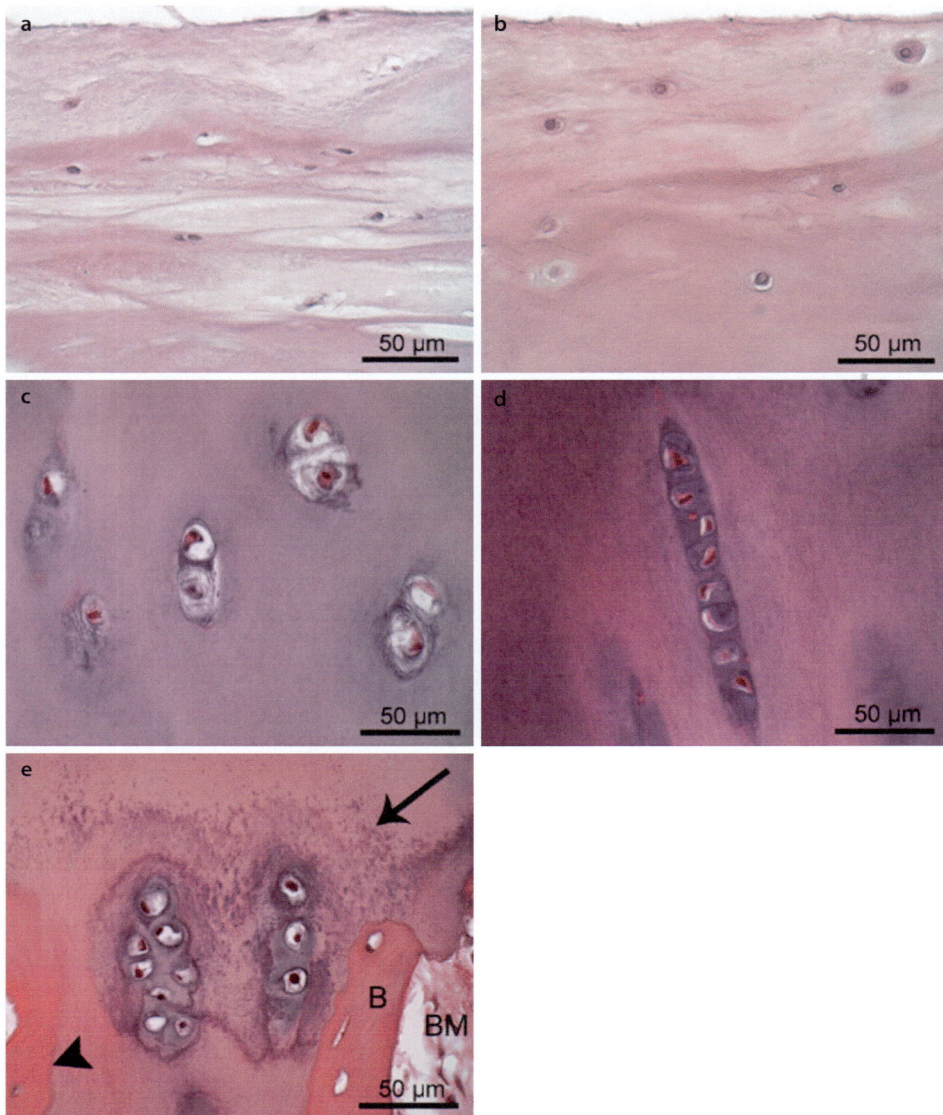

Fig. 9.4 Morphology and arrangement of the chondrocytes and chondrons in the four regions of articular cartilage: In the superficial layer, they are either **a** elongated and aligned parallel to the joint surface or **b** spherical, but in both cases they are individually distributed in the matrix. **c** In the middle zone, chondrocytes are spherical and arranged in pairs **d** while they form stacks of several cells (multichondron) in the deep zone. **e** Multichondrons are also found in the calcified zone, but cells are arrange in three-dimensional groups. Note the tidemark at the interface of calcified and non-calcified cartilage (*arrow*) and the cement line separating the cartilage from the bone (*arrowhead*). *B* bone, *BM* bone marrow cavity (haematoxylin and eosin-stained human femur condylus) (© Georg Thieme Verlag KG)

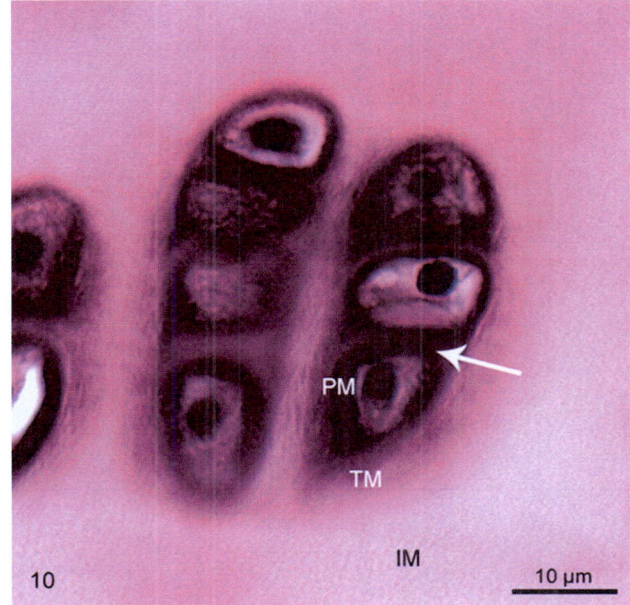

Fig. 9.5 Haematoxylin and eosin-stained cartilage of the rabbit with a paired multichondron, each surrounded by a territorial matrix. Each individual cell is separately (*arrow*) enclosed by a pericellular matrix (*PM*) forming a dense intensively stained rim. *TM* territorial matrix, *IM* interterritorial matrix (© Georg Thieme Verlag KG)

This zonal pattern is consistent in most adult articular cartilages but can show variation in the relative proportions of each layer, particularly in the middle and deep layers. The surface layer shows variations in the presence of collagen type I and the shape of the chondrocytes (more or less elongated). The calcified cartilage shows only minor variations, but the tide mark shows changes in load conditions by way of remodelled or multiple replicates of the tide mark.

This four-zone concept is an artificial zonation corresponding to changing characteristics between surface and depth in articular cartilage. In any case, three structural delineations exist and are of importance for the mechanobiological and physiological functionality of the tissue. The *tide mark* strengthens the collagen network at the delicate interface of calcified and non-calcified cartilage where stress accumulates while loading. Structurally, the collagen fibrils of the upper and the lower region interdigitate with each other, leading to an especially dense matrix that appears as intensively stained stripes in histology. Multiple tide marks may arise when load conditions change. The *cement line* is less conspicuous than the tide mark and the abrupt but tight interface between bone and calcified cartilage. It is undulated, allowing for intense cohesion. Remodelling of the subchondral plate is done by osteoclasts. Invasion of the bone marrow cavity and blood vessel invasion into the cartilage may be the consequence. While the cement line and calcified cartilage form the barrier towards the bone, the upper surface of the cartilage is delineated by a collagen sheet sealed with a greasy substance, the glycoprotein *lubricin*. Apart from the molecule itself, lubricin also contains fatty acids and hyaluronan and is conjointly produced by the cartilage and synovial membrane. This layer is responsible for the gliding properties and shields macromolecules and cells. When this layer is damaged, it frequently marks the start of degradation.

These three boundaries play an important role in the isolation of the tissue from the synovia or bone marrow, its macromolecules, enzymes and cells. This is one reason for the persistence of the tissue in its specific native state but also its fragility and its inability to regenerate cartilage defects.

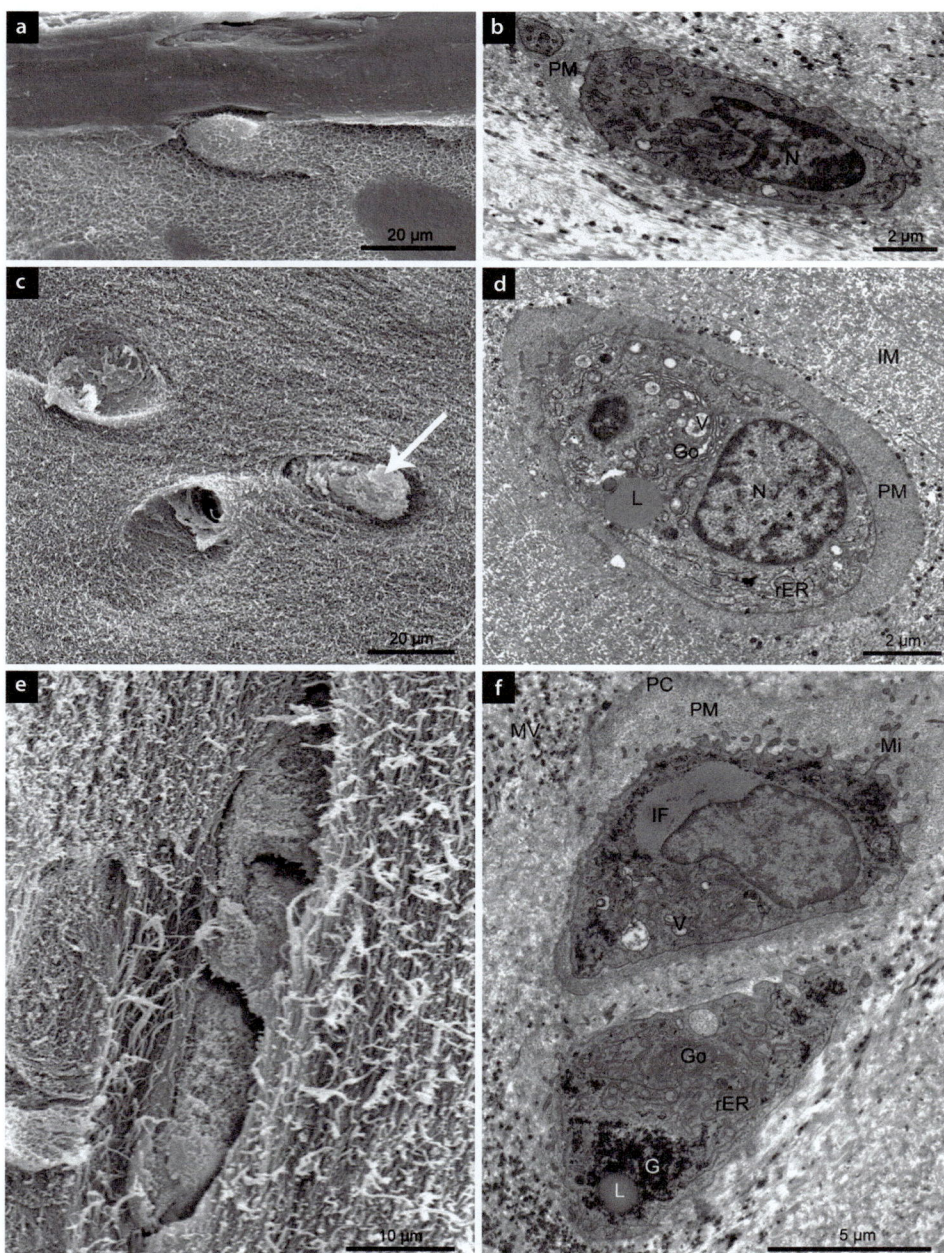

Fig. 9.6 Scanning (*left row*) and transmission (*right row*) electron microscopy of chondrocytes and their matrix environment. **a, b** Elongated cells in the superficial cartilage surrounded by little pericellular matrix (*PM*) and embedded in interterritorial matrix (*IM*). **c, d** Spherical cell (*arrow*) in the transition zone with a thick homogenous rim of pericellular matrix (*PM*), and intracellular lipid droplets (*L*). **e, f** Multi-chondron of three **e** and two **f** chondrocytes stacked one above the other and surrounded by a joint pericellular matrix (*PM*) that reveals a cap (*PC*) in the TEM image **f**. Lipid droplets (*L*) and glycogen (*G*) and vimentin (*IF*) are visible in the TEM **f**. Note also the Golgi apparatus (*Go*), the rough ER (*rER*), vesicles (*V*) and microvilli (*Mi*) indicated in **d** and **f** (© Georg Thieme Verlag KG)

9.5 Compartmentalisation of the ECM in Relation to the Chondrocytes

The extracellular matrix of articular cartilage does not only change its characteristics from the superficial to the deep zone but has further specialised areas around the cells. The main part of the tissue matrix is the *interterritorial matrix* in which the collagen fibrils are tightly packed together in the form of leaves. These leaves are not discernible in histology or macroscopically in healthy cartilage, but may become discernible in damaged and arthritic cartilage and have, in the past, been called *split lines*. In SEM they can be displayed by cryofracturing since the tissue breaks along these lines of weakness. The exposed leaves appear as flat subcompartments consisting of randomly arranged collagen fibrils (◘ Fig. 9.3). The interterritorial area of the cartilage is rich in proteoglycans, with increasing amounts in the deeper region.

The interterritorial matrix is interrupted at regular distances by nests of cells with a specialised matrix environment surrounding them. This compartmentation of the matrix is also visible macroscopically after special stainings. Moving closer to the cells, the collagen becomes continuously finer and the fibrils are arranged in a more loose and random way. This matrix area is called the *territorial matrix* and surrounds all cells of a cell group (chondron). In histology it stains either brighter or darker than the interterritorial matrix, giving the impression of either a bright or a dark halo around the cells (◘ Figs. 9.2 and 9.5).

Under the transmission electron microscope, a further matrix region becomes visible. It is a narrow rim of fine granular matrix without banded collagen fibrils directly surrounding the cell (◘ Fig. 9.6). This *pericellular matrix* is well demarcated from the surrounding territorial matrix due to a sharp transition from the fine granular to the fibrous matrix. The upper and lower sides, orientated to either the cartilage surface or the deep layers of cartilage, may be especially developed with a denser, more intensively stained upper region (cap) and a peaked, lower opposite pole (tail) with a more continuous transition to the territorial matrix (◘ Fig. 9.6). The presence and dimensions of the pericellular and territorial matrix increase from the superficial to the deep and calcified zone. This is probably due to the effects that compression and hypoxia and the consecutive decreasing perfusion and nutrient supply have on the matrix. The detachment of the cells and their subsequent shrinkage during fixation or the extraction of the pericellular matrix during dehydration leads to an artificial gap between the cells and the territorial matrix. This gap is visible in histology and has previously been called a lacuna. However, under natural conditions, chondrocytes have complete contact with the surrounding matrix.

While electron microscopy helps to display structural details down to the ultrastructural level, immunohistochemistry allows the composition of the matrix and the localisation of the components inside the tissue to be identified. The main components of articular cartilage other than water (up to 80%) are collagen type II (2/3 of dry weight) and proteoglycans (1/3 of dry weight). Both are increased in the deeper areas closer to the bone. Minor components which are also part of the main matrix include the fibrous collagen type XI and the linking collagen type IX, which help form the collagen fibrils, glycosaminoglycans (e.g., chondroitin sulphate, keratan sulphate) and non-collagenous proteins like cartilage oligomer matrix protein (COMP) and matrilins. Some matrix components are found in a specific localisation in the tissue, such as the collagen type I that forms the fibrillary collagen in the superficial zone instead of collagen type II, at least in specific joint areas. The anchoring collagen type VI is accumulated in the pericellular matrix, which is also the case for hyaluronan. Chondroitin sulphate appears in the pericellular as well as territorial matrix.

Despite some specific substances of hyaline cartilage (COMP) having been identified in the last decades, collagen type II and aggrecan remain the most specific and reliable markers for differentiated hyaline cartilage and are used for immunological, biochemical and molecular biological phenotype determination. Alcian blue or thionin staining are the most common histochemical methods to display the glycosaminoglycan (GAG) distribution in tissue sections. Immunohistochemistry allows the identification of particular subclasses of GAGs (chondroitin or keratin sulphate). Most typically, collagen type II is tested versus collagen type I in order to identify the chondrogenic differentiation of a tissue. Ultrastructural studies using cationic electron dense stains allow, especially in transmission electron microscopy, the proteoglycans in the matrix to be preserved and detected (◘ Fig. 9.1c).

9.6 The Chondrocyte and its Pericellular Microenvironment

Articular cartilage comprises a small, but critical, population of chondrocytes embedded within the prominent extracellular matrix. The chondrocytes and their pericellular and territorial matrix are conjointly defined as *chondrons*. It is the functional unit of the cells and their immediate surroundings, providing protection against mechanical loading but also serving as a communication and sensing region in which the cells detect mechanobiological changes. As it is the contact site to the cell surface, it is also the interchange area for nutrients and metabolites and the central area of turnover of matrix components. The cells take up extracellularly degraded matrix components from this space and secrete newly synthesised matrix which is then passively distributed into the territorial and interterritorial matrix. Chondrocytes spread microvilli through the pericellular matrix and, in the calcified cartilage, shed matrix vesicles to initiate the formation of hydroxyl apatite crystals. Another single protrusion each chondrocyte has is the primary cilium. It is capable of sensing mechanical conditions and initiating signalling cascades. These cilia rise from a pocket in the cytoplasm (invagination of the plasma membrane) and leave the level of the cell surface at a slightly inclined angle, stretching into the pericellular matrix (◘ Fig. 9.7). Primary cilia have a conspicuous appearance, and in chondrocytes, the shaft extending into the extracellular space consists of a membrane-bound ciliary axoneme, which has a characteristic 9 + 0 microtubular doublet pattern. Intracellularly, the axonemal microtubule end at a diplosomal basal body that is attached distally to the cell membrane by alar sheets and has basal feet.

Intracellularly, chondrocytes bear characteristics related to the physiological, partially anaerobe conditions in this non-vascularised and dense tissue. They have lipid droplets and glycogen accumulations inside the cytoplasm. And, probably in relation to the mechanical load they are exposed to, they contain bundles of intermediary filaments (vimentin), either arranged in whirls in the cytoplasm or alongside the nucleus.

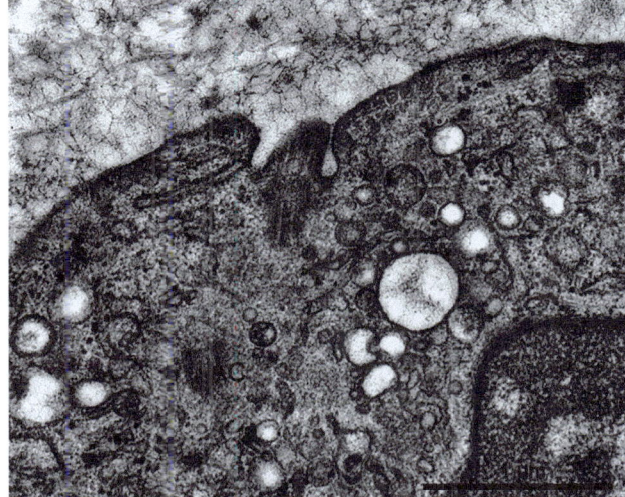

Fig. 9.7 The basal area of a shaft of a primary cilium extending from a depression in the surface of a chondrocyte. The striated structure refers to microtubuli. Note the accessory centriole (*AC*) deeper in the cytoplasm (© Georg Thieme Verlag KG)

9.7 Articular Cartilage Structure Under Load

Articular cartilage represents structurally and functionally differentiated zones of matrices and cells that act synergistically to provide an integrated biological hydroelastic suspension system able to absorb, redistribute and transmit functional compressive forces across articulating joints. The efficacy of this system is primarily dependent on the variation in stiffness, compliance and compressibility of the collagen and proteoglycan components within the tissue region and the metabolic response of the chondrocyte to mechanical, physicochemical and osmotic changes in the articular cartilage matrix, either during physiological function or after damage.

Critical experiments on the tensile properties, compressive stiffness and physicochemical parameters of the matrix indicate the functional integrity of articular cartilage is dependent on (1) the entrapment and confinement of proteoglycans, which act as hydrodynamic substance within the collagen fibre network, and (2) the low hydraulic permeability and high swelling tendency of the hydrophilic proteoglycan component. The structure of this collagen network suggests that the radial collagen alignment of the deeper layers functions as a hydroelastic shock absorber and thus withstands compression, while the tangential collagen fibre sheets of the superficial layer form a tension-resisting diaphragm over the surface of articular cartilage.

At rest, the proteoglycans are maximally hydrated but are limited to around 20% of this potential by the inextensibility of the radial collagen network, which experiences a net tensile load at rest. As load is applied, there is an instantaneous elastic deformation of the matrix at constant volume owing to the incompressibility of the proteoglycan-bound water immobilised by the collagen network. This momentary event is followed by cartilage 'creep', during which time the volume of compressed cartilage decreases as water is forced from the hydrated proteoglycan domains of compressed regions into uncompressed regions of the matrix. A new resting equilibrium is established as the osmotic potential of the proteoglycan gel reaches a pressure comparable to that of the applied load. When load is removed, osmotic pressure exceeds applied pressure and the proteoglycans imbibe water, albeit at a slower rate than that at which it was expressed.

During loading, chondrons are compressed vertically and deformed laterally, but their integrity is maintained by the resistance of the capsule and an adaptive water loss from the pericellular matrix proteoglycans that would dampen the compressive load and provide hydraulic protection for the chondrocyte. Intermittent 'pumping' of the chondron during dynamic loading could also generate microcirculatory currents in the region of the chondron, which might be important for diffusion or the transport of cell metabolites and newly synthesised matrix components into the adjacent matrices.

9.8 Morphological Methods for Cartilage Research

Histology and electron microscopy are the classical methods that previously allowed the characterisation of articular cartilage in detail, and both are still used for cartilage research as morphological gold standard tools. Interference contrast and polarised light microscopy techniques are two of the additional light microscopic tools used to study collagen organisation. Element analysis methods of electron microscopy allow analysis of the composition of tissues. Beyond that, more advanced methods have been implemented and include RAMAN spectroscopy, magnetic resonance imaging (MRI) and computer tomography (CT). These methods may not improve the structural resolution of the classical methods in its current technical state; on the contrary, their output can often first be explained by classical images. However, their advantages are their rapidity, the possibility of untreated (native) sample analysis, *in vivo* applications and 3D observations. Once the techniques and protocols are developed further, these methods will in future contribute to research activities in cartilage regeneration and osteoarthritis research.

> **Take-Home Messages**
> - Hyaline cartilage consists to 90% of extracellular matrix which composes mainly of collagen type II and proteoglycan.
> - Due to the huge amount of proteoglycan that retains water by its negative charges, the matrix is highly hydrated.
> - Articular cartilage is organised in four zones characterised by their matrix organisation and cell characteristics.
> - Chondrocytes are the only cell type in articular cartilage and live under hypoxic conditions and are able to sense and react to changing mechanic conditions.

Suggested Reading

1. Brama PA, Holopainen J, van Weeren PR, Firth EC, Helminen HJ, Hyttinen MM. Effect of loading on the organization of the collagen fibril network in juvenile equine articular cartilage. J Orthop Res. 2009 Sep;27(9):1226–34. doi:10.1002/jor.20866.
2. Goldring MB. Chondrogenesis, chondrocyte differentiation, and articular cartilage metabolism in health and osteoarthritis. Ther Adv Musculoskelet Dis. 2012;4(4):269–85. doi:10.1177/1759720X12448454. PMCID: PMC3403254.

3. Gómez-Camarillo MA, Kouri JB. Ontogeny of rat chondrocyte proliferation: studies in embryo, adult and osteoarthritic (OA) cartilage. Cell Res. 2005;15:99–104. doi:10.1038/sj.cr.7290273.
4. Link TM, Neumann J, Li X. Prestructural cartilage assessment using MRI. J Magn Reson Imaging. 2017;45:949–65. doi:10.1002/jmri.25554.
5. Salinas D, Minor CA, Carlson RP, McCutchen CN, Mumey BM, June RK. Combining targeted metabolomic data with a model of glucose metabolism: toward progress in chondrocyte mechanotransduction. PLoS One. 2017;12(1):e0168326. doi:10.1371/journal.pone.0168326.
6. Ugryumova N, Attenburrow DP, Winlove CP, Matcher SJ. The collagen structure of equine articular cartilage, characterized using polarization-sensitive optical coherence tomography. J Phys D Appl Phys. 2005;38:p2612–9.
7. Van Rossom S, Smith CR, Zevenbergen L, Thelen DG, Vanwanseele B, Van Assche D, et al. Knee cartilage thickness, T1ρ and T2 relaxation time are related to articular cartilage loading in healthy adults. PLoS One. 2017;12(1):e0170002. doi:10.1371/journal.pone.0170002.
8. Zhang Z, Jin W, Beckett J, Otto T, Moed B. A proteomic approach for identification and localization of the pericellular components of chondrocytes. Histochem Cell Biol. 2011;136:153–62.

Growth Plate Research

Gabriele Haeusler, Adalbert Raimann, and Monika Egerbacher

10.1 Introduction – 154

10.2 GP Structure and Function – 154
10.2.1 Endochondral Ossification – 154
10.2.2 GP Structure and Function Is Regulated by Distinct Steps of Differentiation and Proliferation – 155
10.2.3 Complex Local Regulation Loops Regulate Proliferation, Differentiation, and Height of the GP – 156
10.2.4 Endocrine Regulation of Growth Is Dominated by the GH-IGF-I Pathway – 157
10.2.5 The Mechanisms by Which Estrogen Regulates GP Maturation and Fusion Is Still Not Fully Understood – 158

10.3 Techniques in GP Research – 158
10.3.1 Histology – 158
10.3.2 Immunohistochemistry – 160
10.3.3 Electron Microscopy – 161
10.3.4 Laser Capture Microdissection (LCM) – 163
10.3.5 Transgenic Animal Models in GP Research – 164
10.3.6 In Vitro Model Systems in GP Research – 165

References – 168

What You Will Learn in This Chapter
Growth in mammals is a unique and fascinating process which takes place in the growth plates (GP) of long bones and vertebrae.

By growing, the organism achieves adult height, a process that starts in the fetus and is completed at puberty. Growth dynamics are characterized by a strong deceleration (the younger the child, the faster the growth velocity); the only period of growth acceleration in postnatal life is driven by sex steroids during the pubertal growth spurt. The endpoint – adult height – is genetically determined with around 180 genes linked to the variation in final height found in genome-wide association studies [30].

In this chapter, we will describe the GP as an organ and summarize current understanding of the mechanisms of longitudinal growth and its regulation at the local level.

10.1 Introduction

Growth is regulated on several levels and research in the field has concentrated for a long time on hypothalamic and pituitary secretion, the impact of sex steroids (endocrine mechanisms), and influence of health and disease. A set of hypotheses, especially concerning the growth hormone (GH) – insulin-like growth factor 1 (IGF-I) axis – has had to be modified over time. Although clinical observations characterized typical phenotypes of growth disorders, data to explain the pathophysiology of growth on the tissue/cellular level have been very difficult to obtain and are restricted to studies in rodents. The first challenge for studying growth is therefore to establish research models that reflect the human condition as closely as possible and critically use what we learn from them to interpret observations of physiological and pathological growth in children.

10.2 GP Structure and Function

10.2.1 Endochondral Ossification

Long bones and the backbone are molded in cartilage formed from embryonic mesenchymal stem cells that proliferate, condense, and differentiate into chondrocytes. Chondrocytes in the center of the prospective bone become hypertrophic and produce collagen type X. While hypertrophic chondrocytes direct mineralization of the matrix, osteoblasts formed in the perichondrium create a bone collar around the mid-shaft. After completing their task, chondrocytes become apoptotic leaving behind a mineralized scaffold ready for invasion by blood vessels, osteoclasts, and bone marrow from the bone collar.

The mid-shaft ossification center (*primary center of ossification*) gradually expands toward the distal areas. *Secondary centers of ossification* appear in the epiphysis during the late fetal period. The GP, which is responsible for longitudinal growth, comprises the region between the centers (◘ Fig. 10.1). Here chondrocytes continue to proliferate until the centers meet at skeletal maturity and the GP is eradicated.

Although initiated in the fetus, the major events in endochondral bone formation critical for the longitudinal growth and mechanical stability of a healthy skeleton take place between birth and cessation of growth after puberty.

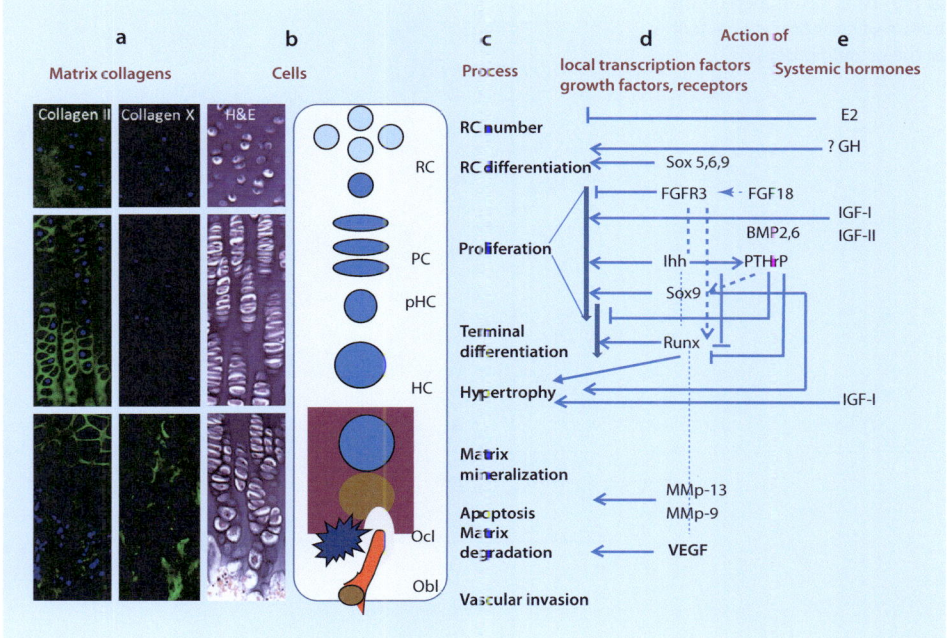

Fig. 10.1 Structure and regulation of the GP. **a, b:** Chondrocytes in the GP are organized in zones, where cell shape and structure indicate their differing functions. See also Fig. 10.3. **b:** Hematoxylin and eosin (H&E)-stained formalin-fixed GP from the proximal tibia of a 25-day-old piglet. **a:** Collagen type II is present throughout the GP, whereas hypertrophic chondrocytes synthesize collagen type X. *RC* reserve zone chondrocyte, *PC* proliferating chondrocyte, *pHC* prehypertrophic chondrocyte, *HC* hypertrophic chondrocyte, *Ocl* osteoclast, *Obl* osteoblast. For details, see text

10.2.2 GP Structure and Function Is Regulated by Distinct Steps of Differentiation and Proliferation (Fig. 10.1c)

Germinal chondrocytes comprising the reserve, or resting, zone within the GP divide infrequently but release their daughter cells into the adjacent proliferating cell zone [19]. The chondrocytes in this zone form a columnar region in which rapid cell division promotes longitudinal bone growth. Ultimately, the chondrocytes exhaust their ability to divide and enter a state of hypertrophy.

Hypertrophy is a crucial driver of longitudinal growth (Fig. 10.2, reviewed in [19]) and an exceedingly efficient and economical process during which chondrocytes undergo great enlargement [5]. Recent studies [8] have shown that chondrocytes pass through three phases of volume increase during enlargement. In the first phase, they produce dry mass and take up a proportional amount of fluid. In the second, fluid is taken up in large amounts disproportional to dry mass production causing a massive swelling of the cells as the fluid dilutes the dry mass density. During the third phase, these now low-density cells continue to enlarge but with a proportional increase in dry mass and fluid again. While variations in the time period of the final phase are important in determining the rate of elongation of GPs, the enormous fluid increase in the second phase is an extraordinary dynamic for enlarging cells and regulating growth rate, considering how critical strict fluid regulation is known to be in the regulation of cell volume.

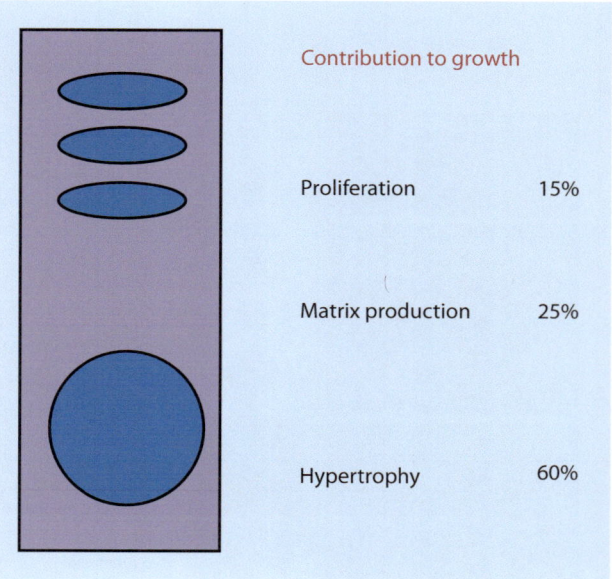

■ **Fig. 10.2** Estimated contribution of chondrocytes' activities to the growth process

Hypertrophic chondrocytes induce mineralization of their surrounding matrix creating a scaffold. Once the cartilage matrix calcifies, terminally differentiated hypertrophic chondrocytes die and disappear. Transverse cartilage septa are resorbed by chondroclasts and blood vessels carrying bone-marrow stromal cells invade the space left behind. Osteoblasts differentiated from the stroma cells subsequently deposit trabecular, or spongy bone on the vertical septa.

While on the cellular level, chondrocytes in the GP affect longitudinal growth by dynamically adding new tissue to bones via a tightly regulated sequence of differentiation, proliferation, and hypertrophy, the height of the plate remains fairly constant to ensure the developing bone is mechanically stable [19].

10.2.3 Complex Local Regulation Loops Regulate Proliferation, Differentiation, and Height of the GP (■ Fig. 10.1d)

Studies in mouse models, in vitro cultures, and skeletal diseases in humans have revealed an ensemble of transcription factors, soluble growth factors, receptors, and extracellular matrix molecules that modulate chondrocyte differentiation and proliferation [4, 28, 36]. The following paragraphs highlight the most important factors and receptors currently known to interact during chondrogenesis.

The transcription factor sex-determining region Y-box 9 (SOX9) in particular has been shown to have critical functions in chondrogenesis (reviewed in [9]). It is expressed in chondroprogenitors and is the principal promoter of chondrocyte proliferation in the early stages of cartilage development. SOX9 activates the chondrocyte-specific genes *COL2A1*, *COL11A2*, and, in concert with SOX5 and SOX6, the chondrocyte-specific enhancers in these genes. Recent work has shown that SOX9 is necessary at many stages in the chondrocyte differentiation pathway to specify and maintain the lineage choice of the cells and activate stage-specific markers (collagen type X).

Fibroblastic growth factors (FGFs) are a large and important family of signaling molecules that act through one or more of four receptors (reviewed in [41]). Fibroblast growth factor receptor 3 (FGFR3), the preferred ligand of FGF18, arrests chondrocyte maturation to maintain a pool of active proliferating chondrocytes. This negative effect of FGFR3 on chondrocyte proliferation is mediated by signaling within the chondrocytes and indirectly by the expression of the Indian hedgehog (IHH), parathyroid hormone-related protein (PTHrP), and bone morphogenetic protein (BMP) signaling pathways. Activating mutations of FGFR3 result in the most prevalent short-limbed dwarfism, achondroplasia, and a number of other skeletal disorders [38].

The PTHrP receptor (PTHrPR), which is activated through the IHH-PTHrP regulatory network, is also important for controlling the rate of bone growth through a negative feedback mechanism. In this loop, mature prehypertrophic chondrocytes secrete the signaling molecule IHH which upregulates PTHrP in the perichondrium and chondrocytes at the articular surface. The PTHrPR on the proliferating cells are thus activated to delay hypertrophy, restricting the number of cells exiting the proliferating zone. In turn, PTHrP's activity is modulated by increased phosphorylation of SOX9 and suppression of runt-related transcription factor 2 (RUNX2), a transcription factor that promotes chondrocyte hypertrophy (for review, see [27, 29]).

BMPs, which belong to the transforming growth factor beta (TGFβ) superfamily, are another set of signaling growth factors critical for limb growth. Their underlying molecular mechanism is not fully understood, but they are believed to regulate chondrocyte differentiation through a complex interaction with the IHH-PTHrP network. There is also evidence that they can act independently of the IHH-PTHrP network (reviewed in [54]).

Little is known about the molecular mechanisms that control chondrocyte enlargement, or the regulation of final cell size. SOX9 and RUNX2 regulate transcription of collagen type X, matrix metalloproteinase 13 (MMP-13), vascular endothelial growth factor (VEGF), and IHH genes (reviewed in [50]) and activate further transcription factors involved in hypertrophy and differentiation such as activating transcription factor 6a (ATF6a) [16].

The cellular events outlined in the above paragraphs require matrix degradation. Enzymes from the matrix metalloproteinase family are regarded as the most important degraders of extracellular matrix (ECM) proteins and are known to play a part in the regulation of cell migration. Their importance in GP biology has been highlighted by distinct ossification defects and morphological GP abnormalities described in knockout mouse models (reviewed in [42]). Furthermore, the relevant MMPs have been found in postnatal human GPs [17]. MMP-13, which is produced by hypertrophic chondrocytes, cleaves collagen type II, thereby facilitating chondrocyte hypertrophy. It acts in concert with MMP-9, produced by osteoclasts, to regulate resorption at the chondro-osseous junction. VEGF, also produced by osteoclasts, appears to be the factor responsible for ingrowth of the vascular progenitor cells in the GP and is regulated by the MAPK pathway [32].

10.2.4 Endocrine Regulation of Growth Is Dominated by the GH-IGF-I Pathway (◘ Fig. 10.1e)

Although several hypotheses on the regulation of bone growth by the GH-IGF-I axis have had to be adapted over time [26, 31], growth disorders involving GH, GH receptor (GHR), IGF-I, IGF-I receptor (IGF-1R), GH signaling, and the acid labile subunit (ALS) point to the significance of these components in endocrine growth regulation [55]. Data on a direct role of GH in differentiation of reserve cell chondrocytes have shown partly contradictory results during the last decades, although activity of the GH signaling cascade in reserve cells has been demonstrated [14]. Recent data indicate that suppressor of cytokine signaling 2 (SOCS2) modulates GH action locally at the GP via a negative signaling loop [1]. There is evidence for an effect of local IGF-I on chondrocyte proliferation, not necessarily produced by GP chondrocytes, but the major local effect of IGF-I seems to be on chondrocyte hypertrophy [43, 46, 52]. IGF-II, which for a long time was only regarded to regulate growth prenatally, was found to be produced locally within the GP in significantly greater amounts than IGF-I [46]. Taken together, although there has been considerable progress in unraveling the action of the GH-IGF-I axis on GP chondrocytes, differences in the phenotypes of rodent models and human growth disorders are as difficult to integrate as are in vitro data on the effect of dependent and independent actions of GH and IGF-I on chondrocyte function.

10.2.5 The Mechanisms by Which Estrogen Regulates GP Maturation and Fusion Is Still Not Fully Understood

In humans but not rodents, GP fuses after puberty and this process is driven by estradiol (E2) in both sexes. Both estrogen receptors alpha (ERα) and beta (ERß) as well as the G-protein-coupled receptor 30 (GPR30) have been shown to be present in all zones of the GP [6, 10], but the main target cell for estrogen action seems to be the reserve zone, where E2 was recently reported to cause irreversible depletion of progenitor cells [39].

10.3 Techniques in GP Research

10.3.1 Histology

Tissue Sampling

The GPs for investigation are usually taken from the distal femur and the proximal tibia (◘ Fig. 10.3). The epiphysis is cut in half using a scalpel or a larger knife. The epiphysis is then broken off the shaft of the bone, whereby the GP remains on the epiphysis. The breakpoint is in the ossification front more or less just below the hypertrophic zone of the GP. The GP is then divided into pieces, which are collected into vessels for conservation or fixation and storage.

As cartilage is a complex tissue, no single preservation technique is able to optimally fix all the structural components. Therefore, different conservation methods and histological fixatives may be used depending on the subsequent procedures planned (see review [22]). Conventional methods are cryo-conservation, formaldehyde, and glutaraldehyde fixation, and more specific ones are methacarn fixation and HOPE® (◘ Fig. 10.4, hepesglutamic acid buffer-mediated organic solvent protection effect) fixation [45], as well as high-pressure freezing, freeze substitution, and low-temperature embedding [20, 49]. However, highly sophisticated preparation methods limit the sample size and therefore are not suitable for microscopic anatomy studies of the GP.

For cryo-conservation, pieces of GP are placed in cryo-tubes, shock-frozen in liquid nitrogen, and then stored at −80 °C. For cryo-sectioning, the frozen GP is adhered to the specimen holder and completely embedded in Tissue Tek® Optimum Cutting Temperature (OCT™)

Growth Plate Research

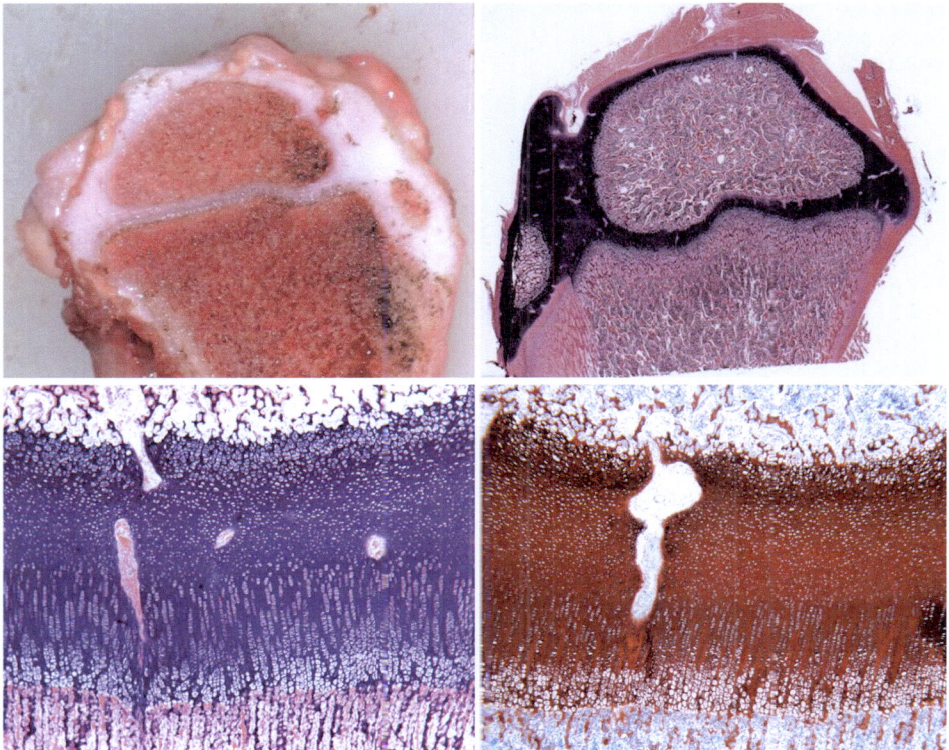

Fig. 10.3 Tissue preparation. Proximal tibia of a 25-day-old piglet with GP immediately after dissection, overview of H&E stained section, and detail of the GP stained with H&E and Safran n O

compound. The ideal cutting temperature is −15 °C. For conventional formalin fixation, GP pieces are fixed in 4% buffered formaldehyde for at least 24 h before paraffin embedding.

Fixation using methacarn solution, consisting of one part glacial acetic acid, three parts chloroform, and six parts absolute methanol, should not exceed 24 h. Samples fixed in formalin and methacarn are finally embedded in 56 °C melting paraffin. Methacarn as a non-cross-linking protein-precipitating fixative usually gives superior immunohistochemical results and is particularly suitable for analysis of molecular events at the mRNA level in paraffin-embedded tissue, especially in conjunction with laser capture microdissection (LCM).

The HOPE® fixation is carried out according to a detailed manufacturer's protocol. Briefly, pieces of tissue are placed in 5 ml of HOPE®-I reagent and samples are stored on crushed ice in a cool room at +4 °C for up to 72 h. After this fixation period, the HOPE®-I reagent is removed and replaced by 5 ml cold acetone and 5 μl HOPE®-II reagent This mixture is incubated for 2 h on crushed ice at +4 °C, followed by three changes of fresh acetone with 2-h incubation after each step. At the end of this fixation process, the samples are embedded in paraffin that melts at 50 °C. Finished paraffin blocks with the HOPE®-fixed specimens have to be stored at +4 °C, while formalin- and methacarn-fixed samples can be stored at room temperature. HOPE®-fixed sections provide excellent preservation of proteins and antigenic structures for differential analysis by immunohistochemical and/or enzyme histochemical techniques. In addition, the most remarkable feature is the extremely low degradation of nucleic acids (DNA and RNA) yielding good results using in situ hybridization techniques.

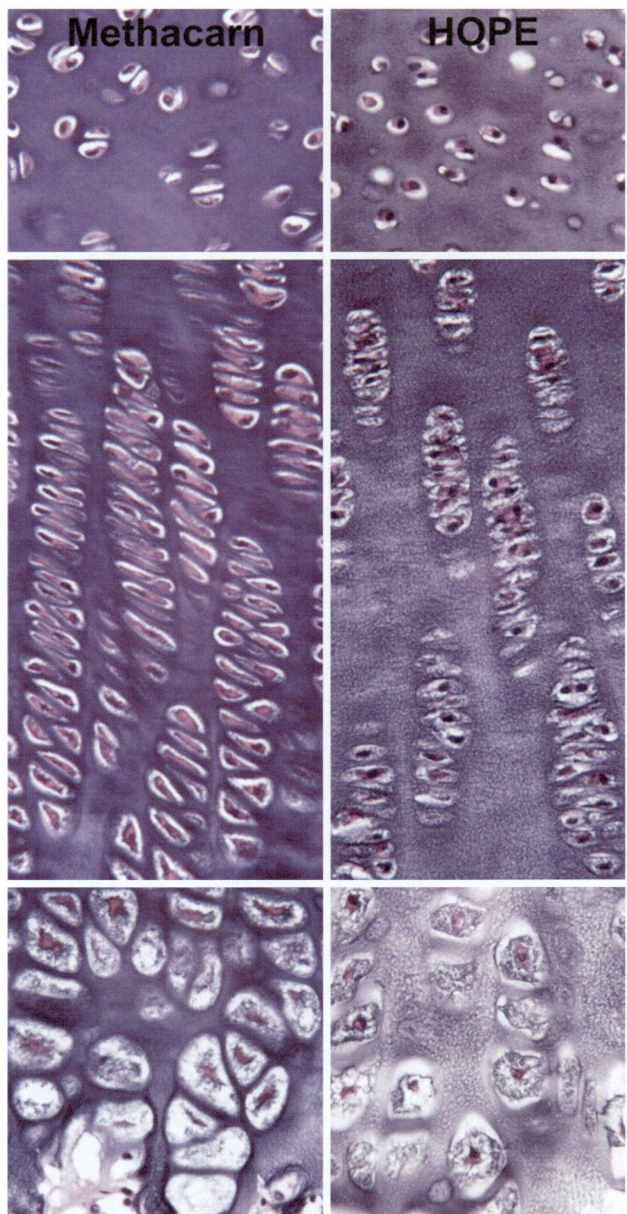

■ **Fig. 10.4** Effect of different fixation protocols on the preservation of GP chondrocytes

10.3.2 Immunohistochemistry

Slides should be coated with 3-aminopropyltriethoxysilane (APES) for immunohistochemical staining of GP cartilage sections. Briefly, immerse slides in a freshly prepared 2% APES-methanol solution (e.g., Sigma A3648) for 5 min, followed by 100% methanol for 5 min. Wash twice in distilled water, dry overnight at 37 °C, fix slides in 3% glutaraldehyde in aqua bidest, dry overnight at 37 °C (or at room temperature), and store coated slides at room temperature.

Different pretreatments and antigen retrieval procedures are required before immunostaining for detection of antigens. Antibodies against matrix proteins like collagen types I, II, IV, and X, laminin, and fibronectin can be used after enzymatic pretreatment.
- Collagen type II and X sections are digested with 0.01% hyaluronidase in phosphate-buffered saline (PBS), pH 6.7, at 37 °C for 4 h and subsequently with pepsin (1 mg/ml 0.1 N HCl) at 37 °C for 30 min.
- Collagen type I and aggrecan sections are digested with 0.1% pepsin in 0.5 M acetic acid at 37 °C for 2 h.
- Collagen type IV, laminin, and fibronectin sections are digested with protease (0.1% protease in 0.05 M Tris-HCl buffer, pH 7.2) at 37 °C for 2 min [18].
- MMP-2, MMP-1, and MMP-14 sections are digested with hyaluronidase (25 mg/50 ml) in PBS, pH 6.7, at 37 °C for 1 h.
- MMP-1, MMP-13, MMP-16, and tissue inhibitors of metalloproteinase-2 (TIMP-2) require antigen retrieval with target unmasking fluid (TUF) at 90 °C for 10 min and subsequent cooling for 15 min.
- TIMP-1, transglutaminase II, and VEGF require antigen retrieval with 0.01 M citrate buffer, pH 6.0, at 90–95 °C [17] for 20 min.

Immunostaining of other proteins, like transmembrane cell-adhesion proteins such as integrins, require pretreatment with collagenase type IA (1 mg/ml PBS, at room temperature for 10 min). Cryo-stat sections have to be used for detection of integrin αV and $\alpha 6$ subunits [18]. Detection of IGF-1R and IGF-2R on GP tissue sections requires pretreatment in 0.001 M EDTA at 96 °C [2] for 20 min and of ERα and ERß in citrate buffer, pH 6.0, at 96 °C [10] for 20 min. No pretreatment is necessary for IGF-I and MGF [47]; however, IGF-II detection requires pretreatment in 0.001 M EDTA, pH 8.0, at 96 °C for 20 min [46].

10.3.3 Electron Microscopy

Specimens are fixed in 3% glutaraldehyde in 0.1 M Sörensen phosphate buffer, pH 7.4, for transmission electron microscopy (Fig. 10.5), using a standard protocol. After washing in phosphate buffer, the specimens are postfixed in a 1% solution of osmium tetroxide and then washed in phosphate buffer again. They are dehydrated in a series of graded ethanol solutions (70, 80, 96, and 100%) and subsequently infiltrated with propylene oxide, followed by increasing ratios of epoxy resin-propylene oxide (1:1, 3:1) and finally pure resin. After an additional change, the resin is polymerized a 60 °C. Semithin sections are cut at 0.8 µm and stained with toluidine blue. Ultrathin sections are cut at 70 nm, mounted on copper grids, and stained with uranyl acetate and lead citrate. This classic aldehyde-based preservation method goes along with cell shrinkage, which is most prominent in hypertrophic chondrocytes and loss of matrix proteins [22]. This effect can be circumvented by addition of cationic dyes, such as toluidine blue, alcian blue, or ruthenium hexamine trichloride (RHT) [21].

Hyper-osmolar fixative solution used in standard protocols is another reason for extensive chondrocyte shrinkage. Therefore, osmotically corrected solutions close to the normal osmotic status in the ECM of the GP (~280 mOsm) are recommended for studies on volume changes within the GP zones and their contribution to bone growth to prevent cell shrinkage [5, 34].

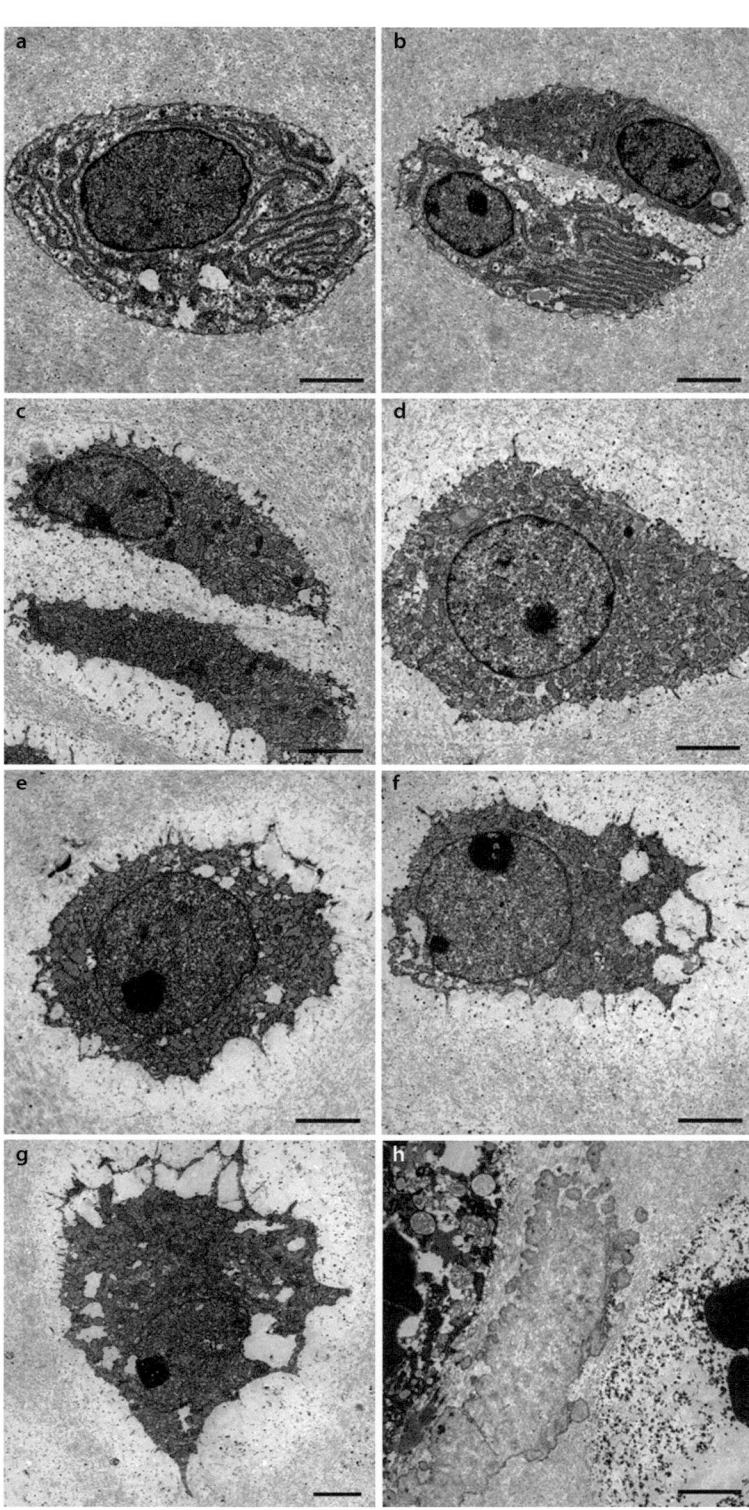

10.3.4 Laser Capture Microdissection (LCM)

Gene expression in chondrocytes in different zones of the GP can be analyzed by LCM and subsequent reverse transcription polymerase chain reaction (RT-PCR) [24, 48]. Pieces of GP are either directly snap frozen and sectioned or conserved in RNase-free tubes containing RNAlater® and handled like the cryo-samples above. If no fresh tissue is available or paraffin-embedded GPs are desired, results are better when samples are fixed in methacarn or HOPE® rather than in formalin. Before use, the chamber of the cryo-tome has to be cleaned with 70% alcohol and the removable parts and instruments with RNase ZAP®, and a new section of the blade or a completely new blade should be used for every new sample.

The sections for LCM can be stained with the Arcturus HistoGene Frozen Section Staining Kit, or left unstained. Unstained frozen sections have to be dipped in ice-cold RNase-free water for a few seconds to remove the OCT™ embedding medium. 6-µm-thick cryo-sections are cut at −15 °C and 3–4 sections mounted on polyethylene naphthalate (PEN) membrane metal-framed LCM slides. Cells of different zones can be pooled and captured on CapSure™ Macro Caps, transferred into lysis buffer, and further processed for RNA extraction (for a detailed description, see (37)).

RNA Isolation and RT-PCR

Gene expression analysis is a crucial tool for functional studies and for characterizing the cells used. RNA is isolated by freezing total tissue explants in liquid nitrogen and homogenizing them with a pestle and mortar or an adequate homogenizer. The method for harvesting monolayer cells does not differ from that used for other cell types. However, non-phosphate-based buffer solutions should be used for washing steps because chondrocytes can be sensitive. Both TRIzol- and columnar-based approaches for RNA isolation can be used for monolayer cells and native tissue. Micro-kits are available for very small samples (e.g., of LCM). TRIzol is preferred for 3D matrices such as alginate to avoid problems with the columnar filters.

Fig. 10.5 Cells of the GP: Resting zone chondrocytes **a** are round to oval shaped with spiky cytoplasmic protrusions extending into the immediate intercellular matrix. Chondrocytes contain a round nucleus, abundant rough endoplasmic reticulum (rER), and a few mitochondria. When cells start to proliferate, the respective daughter cells are still connected by cellular protrusions. They move apart as they begin to secrete ECM into the intercellular space between them **b**. Chondrocytes in the proliferating zone **c** flatten and become spindle shaped as they orient in stacks and columns. Nuclei are positioned at the opposite ends of the consecutive cells and the cytoplasm is filled with rER. Again, thin protrusions connect the cells to their surrounding matrix. The most prominent change at the distal end of each column is an increase in size and change in shape as cells become round again **d**. The change from prehypertrophic to hypertrophic cells is a gradual one, so that it is tempting to describe an early prehypertrophic **d** and a late prehypertrophic cell **e** within the GP: From chondrocytes that completely fill the lacuna to chondrocytes that retract and show more prominent cytoplasmic processes. Cells show extended and swollen rER within the cytoplasm. Hypertrophy is characterized by a volume increase that leads to cell shrinkage in standard processed tissue, producing long cytoplasmic protrusions which extend to the wall of the lacunae. Along with cellular indentations, they give the cells a porous appearance **f, g**. Hypertrophic chondrocytes have an eccentrically situated dense nucleus with an irregular outline and cisterns of rER. At the chondroosseous junction **h**, the matrix calcifies in the longitudinal septa that serve as templates for early osteoid deposition (*). Scale bar = 2500 nm

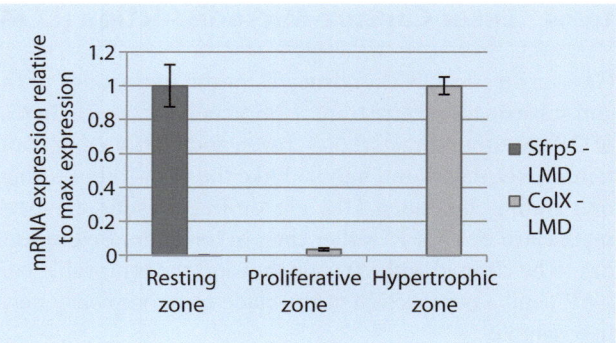

Fig. 10.6 Validation of *Sfrp5* and *Col10A1* as markers of differentiation in porcine LCM samples by RT-PCR (*n* = 3)

The most commonly used reference genes for RT-PCR are *bActin*, *GAPDH*, and *18S*, of which *GAPDH* has shown advantages in cartilaginous tissues, whereas *18S* is sometimes preferred when comparing different tissues. *PPIA*, a non-classic reference gene, has been shown to be an interesting candidate in cell lines and native chondrocytes [57].

Differentiation can be monitored by expression of stage-specific genes. *Col10A1* is the best-established marker and is specific for hypertrophic chondrocytes. Recent microarray data from murine samples revealed new markers such as *Sfrp5* for resting chondrocytes (Fig. 10.6) [35].

10.3.5 Transgenic Animal Models in GP Research

Global Knockouts

Transgenic animal model systems are widely used to study diseases and specific gene functions. Global gene knockouts are frequently used to mimic human diseases and learn about their pathomechanisms. Model animals have been developed for growth-related disorders such as achondroplasia and Stickler syndrome [25, 53]. Other systemic knockouts have improved our knowledge on key regulators of GP physiology such as the GH-IGF-1 axis, IHH-PTHrP, Wnt signaling, and the role of sex steroids. However, taking sex steroids as an example of species-specific differences between human disease and transgenic mice, dissimilar effects of loss of function of the estrogen receptor are evident [7].

Conditional/Inducible Gene Targeting

The term conditional gene targeting covers several techniques that are more specific in altering gene expression than global gene targeting. *Conditional gene targeting* is used to introduce genetic tags such as lox («floxing») into the gene of interest. These tags are recognized by recombinases like Cre. Cre itself is inserted into the genome under the control of a chosen promoter leading to a targeted excision of the «floxed» gene of interest. A different approach of «switching» gene expression is adopted in *inducible gene targeting*: Ligand-binding domains or drug-dependent promoters are inserted and become activated upon administration of non-endogenous substances.

Conditional and inducible gene targeting can be combined to create powerful and flexible model systems for studies focusing on specific tissues such as the GP. The Col2a1-Cre mouse is probably the best-established transgenic model system in chondrocyte research. The Col10a1-Cre mouse, which is used less frequently, is theoretically even more precise for targeting the GP because *Col10A1* in this model is almost exclusively expressed in chondrocytes undergoing hypertrophy [13]. Additionally, «differentiation-specific» Col11alpha2-Cre mice have been developed that express recombinase at the onset of the columnar proliferation stage [23].

Caveats

Interpreting transgenic mouse studies can be challenging, for example, when the results from global and targeted knockout mice conflict or systemic physiological changes are induced that alter gene expression. Most importantly, it should be noted that the term «chondrocyte-specific» is misleading as effects on other cells of mesenchymal lineage such as osteoblasts can be expected. For example, a Col2-Cre IGF-1 knockout often cited for endogenous synthesis of IGF-1 in chondrocytes shows similar Cre expression levels in the bone and in the cartilage [15]. Furthermore, to a certain extent, Cre expression itself alters bone and GP architecture, resulting in a need for Cre +/+ control groups (personal communication Prof. H. Kronenberg).

10.3.6 In Vitro Model Systems in GP Research

As human GP material is not readily available, model systems are usually used to study local mechanisms within the GP including regulation of gene expression and cellular interactions. Three different cell culture approaches are broadly used: primary chondrocytes, differentiated mesenchymal stem cells (MSCs), and cell lines.

Primary Cells

Primary Cell Isolation

Primary chondrocytes can be derived directly from GP cartilage. To ensure a sufficient number of chondrocytes, the cartilage is procured from large prepubertal mammals such as piglets, cattle, or horses. The entire epiphysis is broken off at the ossification front and mechanically loosened from the diaphyseal bone. The GP tissue obtained is incubated in media containing collagenase II, filtered, and centrifuged to collect the cells [46].

Rib cartilage is frequently used when GP cartilage is unavailable or sparse and can be obtained from embryonic or up to day 3 postnatal animals. The thorax is dissected and non-cartilaginous tissues are removed under a microscope. After incubation with antibiotic solution and repeated cleansing, the remaining cartilaginous ribs are digested in collagenase II solution [44].

> **Tip**
>
> It is important in both approaches to ensure that the tissue is thoroughly cleansed to minimize contamination with fibroblasts or bone cells. Furthermore, animals of the same breed, age, sex and strain should be used.

Cell Subfractioning

Techniques to isolate subpopulations have been developed to overcome the cellular inhomogeneity of the GP. Large amounts of chondrocytes can be retrieved using *density gradient centrifugation* after collection of primary chondrocytes [40]. Percoll® dilutions can be made and layered to obtain a discontinuous gradient. The falcons have to be handled and centrifuged with extreme caution. The following densities have been used successfully in porcine cells [47].

<1032 g/ml   mostly prehypertrophic cells
<1049 g/ml   mostly proliferative cells
<1065 g/ml   mostly resting cells
>1065 g/ml   mostly blood cells

Mesenchymal Stem Cells (MSCs)

Using mesenchymal stem cells has the advantages that the cells are obtainable from human material and have high proliferative capacities. Protocols are available for deriving MSCs from fetal and postnatal sources such as the bone marrow, muscle, or periosteum [3]. Chondrogenic differentiation of MSCs is dependent on the presence of a functional extracellular matrix in culture.

Cell Lines

Multiple chondrosarcoma and immortalized chondrocyte cell lines have been developed to bypass the lack of primary cells. As always, cell-line-specific differences to normal cell physiology have to be taken in account.

ATDC5, which is derived from a murine teratocarcinoma, is widely used in cartilage research. The cells can express high levels of *Col10A1* in monolayer culture, making it an attractive system for a spectrum of applications in GP research [56].

The most commonly used human chondrocyte lines, T/C-28 and C-28/I2, feature high proliferation rates, easy handling, and expression of chondrocyte-specific genes such as SOX9. However, their low expression of ECM genes and the absence of *Col10A1* expression limit their use for GP research [11].

Metatarsal Culture

Culturing total metatarsal bone including their GPs preserves the cells' physiological environment and avoids the potential systemic effects of treatments in in vivo models. As murine metatarsals are readily available and easy to handle, this system is frequently used in GP research. Most protocols describe the culture of embryonic metatarsals derived from the rats' middle metatarsal bones. Only embryonic or very early postnatal bones can be used to avoid perfusion issues [37].

Tip	
Caution is required when comparing metatarsal and long bone physiology, as well as embryonic and postnatal physiology.	

Culture Conditions

Chondrocyte Media

Many different protocols are applied for optimal culture media for chondrocytes. Although commercial formulas are available, self-made formulas are more cost-effective and flexible. Typical media are based on DMEM or DMEM-Hams F12F. For induction of differentiation in serum-free media, a spectrum of agents are used to induce differentiation in serum-free media including β-glycerophosphate (bGP), ascorbic acid, insulin transferrin selenium (ITS), TGFβ, dexamethasone, BMPs, and FGFs.

- *Serum-containing differentiation medium:* DMEM with 10% fetal calf serum (FCS), 2 µg/ml amphotericin B, 100 µg/ml gentamycin, 50 µg/ml ascorbic acid phosphate, 5–100 nM insulin, 4 mM glutamine
- *Serum-free medium for treatments* (modified from Loeser and Shanker [33]): Phenol-red free DMEM with 2 mM Na pyruvate, 4 mM glutamine, 0.5 ug/ml transferrin, 2 ng/ml selenious acid, 420 µg/ml BSA, 2.1 µg/ml linoleic acid, 25 mg/ml ascorbic acid, 3150 mg/ml glucose, 0–100 nM insulin

Tips

- The number of infections is often lower with gentamycin/amphotericin than with Pen-Strep.
- The concentration of bGP in differentiation media should not exceed 2 mM to avoid ectopic mineralization.
- Avoid using media containing phenol red during treatments.
- The amount of insulin should be lowered when analyzing GH-IGF-1 effects to avoid saturation of the IGF-1R.
- If trypsinization problems are encountered, try predigestion with media containing collagenase II or Accutase®.
- Ascorbic acid phosphate is often preferred due to its longer half-life time. However, if phosphate load is to be avoided, ascorbic acid should be used and media changed at least every 48 h.

3D Culture Systems

The growth plate contains a complexly orchestrated extracellular matrix. Thus, in vitro culture of chondrocytes faces the challenge of keeping the cells in a physiological, differentiated state despite the loss of the original environment. Chondrocytes alter their expression profile in monolayer culture and undergo a dedifferentiation to a fibroblast-like cell type. Several models for 3D chondrocytes cultures have been developed to preserve a chondrocyte phenotype and allow further differentiation of the cells.

Alginate is a purified algae-derived linear polysaccharide which has been well-established in chondrocyte culture. Both preservation of the growth plate chondrocyte phenotype and redifferentiation in culture have been shown, qualifying this model as an excellent culture system for GP research [2].

Gelatin-coated wells are often used in MSC differentiation, but have also proven to enhance differentiation of primary chondrocytes. Coating with 0.1% gelatin is easy and cost-efficient compared with other systems.

Collagen scaffolds such as gels, membranes, and sponges are common in orthopedic and tissue engineering research [12]. *Other scaffolds* consisting of hyaluronic acid, chitosan, cellulose, agarose, and fibrin glue have been used in tissue engineering.

Micromass cultures allow cells to create an endogenous 3D environment. Initially, a cell suspension is centrifuged to create a pellet. During cultivation in chondrogenic medium, the pellet compacts by organization of the chondrocytes and ECM synthesis. This system has successfully been used in primary cells, mesenchymal stem cells, and cell lines [51].

> **Tips**
> - The shape of some tubes is not optimal for creating stable pellets; different types and vendors should be tested.
> - Alginate can impair cDNA synthesis and polymerase activity; repeated washing steps before cell lysis are essential.

> **Take-Home Messages**
> - Endochondral bone formation is the central mechanism behind longitudinal growth and mechanical stability of a healthy skeleton.
> - Growth plate chondrocyte hypertrophy is a crucial driver of longitudinal growth.
> - Important factors known to regulate the architecture of the growth plate comprise IGF-1, IHH-PTHrP, TGFβ, and SOX9.
> - Estrogens regulate growth plate maturation and fusion by a yet not fully understood mechanism.
> - Due to its complex mechanical features, adapted modes of cutting and preparation are crucial for histologic samples of the growth plate.
> - Conditional and inducible gene targeting revealed new insight into regulatory mechanisms of the growth plate.
> - Common problems in in vitro models of the growth plate are dedifferentiation, lack of appropriate matrix, and species-specific conditions and effects.

References

1. Ahmed SF, Farquharson C. The effect of GH and IGF1 on linear growth and skeletal development and their modulation by SOCS proteins. J Endocrinol. 2010;206:249–59. doi:10.1677/JOE-10-0045.
2. Albrecht C, Helmreich M, Tichy B, et al. Impact of 3D-culture on the expression of differentiation markers and hormone receptors in growth plate chondrocytes as compared to articular chondrocytes. Int J Mol Med. 2009;23(3):347–55. doi:10.3892/ijmm_00000138.
3. Atala A, Lanza RP. Methods of tissue engineering. San Diego: Gulf Professional Publishing; 2002.
4. Ballock RT, O'Keefe RJ. The biology of the growth plate. J Bone Jt Surg. 2003;85:715–26.
5. Bush PG, Hall AC, Macnicol MF. New insights into function of the growth plate clinical observations, chondrocyte enlargement and a possible role for membrane transporters. J Bone Joint Surg Br. 2008;90:1541–7.

6. Chagin AS, Sävendahl L. GPR30 estrogen receptor expression in the growth plate declines as puberty progresses. J Clin Endocrinol Metab. 2007a;92:4873–7. doi:10.1210/jc.2007-0814.
7. Chagin AS, Sävendahl L. Oestrogen receptors and linear bone growth. Acta Paediatr. 2007b;96:1275–9. doi:10.1111/j.1651-2227.2007.00415.x.
8. Cooper KL, Oh S, Sung Y, et al. Multiple phases of chondrocyte enlargement underlie differences in skeletal proportions. Nature. 2013;495:375–8. doi:10.1038/nature11940.
9. Dy P, Wang W, Bhattaram P, et al. Sox9 directs hypertrophic maturation and blocks osteoblast differentiation of growth plate chondrocytes. Dev Cell. 2012;22:597–609. doi:10.1016/j.devcel.2011.12.024.
10. Egerbacher M, Helmreich M, Rossmanith W, Haeusler G. Estrogen receptor-alpha and estrogen receptor-beta are present in the human growth plate in childhood and adolescence, in identical distribution. Horm Res. 2002;58:99–103.
11. Finger F, Schörle C, Zien A, et al. Molecular phenotyping of human chondrocyte cell lines T/C-28a2, T/C-28a4, and C-28/I2. Arthritis Rheum. 2003;48:3395–403. doi:10.1002/art.11341.
12. Gavénis K, Schmidt-Rohlfing B, Mueller-Rath R, et al. In vitro comparison of six different matrix systems for the cultivation of human chondrocytes. Vitro Cell Dev Biol - Anim. 2006;42:159–67. doi:10.1290/0511079.1.
13. Gebhard S, Hattori T, Bauer E, et al. Specific expression of Cre recombinase in hypertrophic cartilage under the control of a BAC-Col10a1 promoter. Matrix Biol. 2008;27:693–9. doi:10.1016/j.matbio.2008.07.001.
14. Gevers EF, Hannah MJ, Waters MJ, Robinson ICAF. Regulation of rapid signal transducer and activator of transcription-5 phosphorylation in the resting cells of the growth plate and in the liver by growth hormone and feeding. Endocrinology. 2009;150:3627–36. doi:10.1210/en.2008-0985.
15. Govoni KE, Lee SK, Chung YS, et al. Disruption of insulin-like growth factor-I expression in type IIalpha1 collagen-expressing cells reduces bone length and width in mice. Physiol Genomics. 2007;30:354–62. doi:10.1152/physiolgenomics.00022.2007.
16. Guo F, Han X, Wu Z, et al. ATF6a, a Runx2-activable transcription factor, is a new regulator of chondrocyte hypertrophy. J Cell Sci. 2016;129:717–28. do :10.1242/jcs.169623.
17. Haeusler G, Walter I, Helmreich M, Egerbacher M. Localization of matrix Metalloproteinases, (MMPs) their tissue inhibitors, and vascular endothelial growth factor (VEGF) in growth plates of children and adolescents indicates a role for MMPs in human postnatal growth and skeletal maturation. Calcif Tissue Int. 2005;76:326–35. doi:10.1007/s00223-004-0161-6.
18. Häusler G, Helmreich M, Marlovits S, Egerbacher M. Integrins and extracellular matrix proteins in the human childhood and adolescent growth plate. Calcif Tissue Int. 2002;71:212–8. doi:10.1007/s00223-001-2033-x.
19. Hunziker EB. Mechanism of longitudinal bone growth and its regulation by growth plate chondrocytes. Microsc Res Tech. 1994;28:505–19.
20. Hunziker EB, Schenk RK. Cartilage ultrastructure after high pressure freezing, freeze substitution, and low temperature embedding. II Intercellular matrix ultrastructure - preservation of proteoglycans in their native state J Cell Biol. 1984;98:277–82.
21. Hunziker EB, Herrmann W, Schenk RK. Ruthenium hexammine trichloride (RHT)-mediated interaction between plasmalemmal components and pericellular matrix proteoglycans is responsible for the preservation of chondrocytic plasma membranes in situ during cartilage fixation. J Histochem Cytochem. 1983;31:717–27.
22. Hunziker EB, Lippuner K, Shintani N. How best to preserve and reveal the structural intricacies of cartilaginous tissue. Matrix Biol. 2014;39C:33–43. doi:10.1016/j.matbio.2014.08.010.
23. Iwai T, Murai J, Yoshikawa H, Tsumaki N. Smad7 inhibits chondrocyte differentiation at Multiple steps during Endochondral bone formation and down-regulates p38 MAPK pathways. J Biol Chem. 2008;283:27154–64. doi:10.1074/jbc.M801175200.
24. Jacquet R, Hillyer J, Landis WJ. Analysis of connective tissues by laser capture microdissection and reverse transcriptase-polymerase chain reaction. Anal Biochem. 2005;337:22–34. doi:10.1016/j.ab.2004.09.033.
25. Kaarniranta K, Ihanamäki T, Sahlman J, et al. A mouse model for Stickler's syndrome: ocular phenotype of mice carrying a targeted heterozygous inactivation of type II (pro)collagen gene (Col2a1). Exp Eye Res. 2006;83:297–303. doi:10.1016/j.exer.2005.11.027.
26. Kaplan SA, Cohen P. REVIEW: the Somatomedin hypothesis 2007: 50 years later. J Clin Endocrinol Metab. 2007;92:4529–35. doi:10.1210/jc.2007-0525.

27. Komori T. Signaling networks in RUNX2-dependent bone development. J Cell Biochem. 2011;112:750–5. doi:10.1002/jcb.22994.
28. Kronenberg HM. Developmental regulation of the growth plate. Nature. 2003;423:332–6.
29. Kronenberg HM. PTHrP and Skeletal Development. Ann N Y Acad Sci. 2006;1068:1–13. doi:10.1196/annals.1346.002.
30. Lango Allen H, Estrada K, Lettre G, et al. Hundreds of variants clustered in genomic loci and biological pathways affect human height. Nature. 2010;467:832–8. doi:10.1038/nature09410.
31. Le Roith D, Bondy C, Yakar S, et al. The somatomedin hypothesis: 2001. Endocr Rev. 2001;22:53–74.
32. Liu ES, Raimann A, Chae BT, et al. c-Raf promotes angiogenesis during normal growth plate maturation. Development. 2016;143:348–55. doi:10.1242/dev.127142.
33. Loeser RF, Shanker G. Autocrine stimulation by insulin-like growth factor 1 and insulin-like growth factor 2 mediates chondrocyte survival in vitro. Arthritis Rheum. 2000;43:1552–9. doi:10.1002/1529-0131(200007)43:7<1552::AID-ANR20>3.0.CO;2-W.
34. Loqman MY, Bush PG, Farquharson C, Hall AC. A cell shrinkage artefact in growth plate chondrocytes with common fixative solutions: importance of fixative osmolarity for maintaining morphology. Eur Cell Mater. 2010;14:214–27.
35. Lui JCK, Andrade AC, Forcinito P, et al. Spatial and temporal regulation of gene expression in the mammalian growth plate. Bone. 2010;46:1380–90. doi:10.1016/j.bone.2010.01.373.
36. Mackie EJ, Tatarczuch L, Mirams M. The skeleton: a multi-functional complex organ. The growth plate chondrocyte and endochondral ossification. J Endocrinol. 2011;211:109–21. doi:10.1530/JOE-11-0048.
37. Mårtensson K, Chrysis D, Sävendahl L. Interleukin-1β and TNF-α act in synergy to inhibit longitudinal growth in Fetal rat metatarsal bones. J Bone Miner Res. 2004;19:1805–12. doi:10.1359/JBMR.040805.
38. Narayana J, Horton WA. FGFR3 biology and skeletal disease. Connect Tissue Res. 2015;56:427–33. doi:10.3109/03008207.2015.1051224.
39. Nilsson O, Weise M, Landman EBM, et al. Evidence that estrogen hastens epiphyseal fusion and cessation of longitudinal bone growth by irreversibly depleting the number of resting zone progenitor cells in female rabbits. Endocrinology. 2014;155:2892–9. doi:10.1210/en.2013-2175.
40. Oberbauer AM, Peng R. Fractionation of growth plate chondrocytes: differential expression of IGF-I and growth hormone and IGF-I receptor mRNA in purified populations. Connect Tissue Res. 1995;31:179–87.
41. Ornitz DM. FGF signaling in the developing endochondral skeleton. Cytokine Growth Factor Rev. 2005;16:205–13. doi:10.1016/j.cytogfr.2005.02.003.
42. Ortega N, Behonick DJ, Werb Z. Matrix remodeling during endochondral ossification. Trends Cell Biol. 2004;14:86–93. doi:10.1016/j.tcb.2003.12.003.
43. Parker EA, Hegde A, Buckley M, et al. Spatial and temporal regulation of GH-IGF-related gene expression in growth plate cartilage. J Endocrinol. 2007;194:31–40. doi:10.1677/JOE-07-0012.
44. Sabbagh Y, Carpenter TO, Demay MB. Hypophosphatemia leads to rickets by impairing caspase-mediated apoptosis of hypertrophic chondrocytes. Proc Natl Acad Sci U S A. 2005;102:9637–42.
45. Scharmer D. Differential gene expression in chondrocytes of the resting, proliferative and hypertrophic zone of the growth plate in the pig at different stages of development [dissertation]. University of Veterinary Medicine, Vienna, Austria 2009.
46. Schlegel W, Halbauer D, Raimann A, et al. IGF expression patterns and regulation in growth plate chondrocytes. Mol Cell Endocrinol. 2010;327:65–71. doi:10.1016/j.mce.2010.06.005.
47. Schlegel W, Raimann A, Halbauer D, et al. Insulin-like growth factor I (IGF-1) Ec/Mechano growth factor – a splice variant of IGF-1 within the growth plate. PLoS One. 2013;8:e76133. doi:10.1371/journal.pone.0076133.
48. Shao YY, Wang L, Hicks DG, Ballock RT. Analysis of gene expression in mineralized skeletal tissues by laser capture microdissection and RT-PCR. Lab Invest. 2006;86:1089–95. doi:10.1038/labinvest.3700459.
49. Studer D, Michel M, Wohlwend M, et al. Vitrification of articular cartilage by high-pressure freezing. J Microsc. 1995;179:321–32.
50. Studer D, Millan C, Oeztuerk E, et al. Molecular and biophysical mechanisms regulating hypertrophic differentiation in chondrocytes and mesenchymal stem cells. Eur Cell Mater. 2012;24:118–35.
51. Tare RS, Howard D, Pound JC, et al. Tissue engineering strategies for cartilage generation—Micromass and three dimensional cultures using human chondrocytes and a continuous cell line. Biochem Biophys Res Commun. 2005;333:609–21. doi:10.1016/j.bbrc.2005.05.117.

52. Wang JIE, ZHOU J, BONDY CA. Igf1 promotes longitudinal bone growth by insulin-like actions augmenting chondrocyte hypertrophy. FASEB J. 1999;13:1985–90.
53. Wang Y, Spatz MK, Kannan K, et al. A mouse model for achondroplasia produced by targeting fibroblast growth factor receptor 3. Proc Natl Acad Sci. 1999;96:4455–60. doi:10.1073/pnas.96.8.4455.
54. Wang W, Rigueur D, Lyons KM. TGFβ signaling in cartilage development and maintenance. Birth Defects Res Part C Embryo Today. 2014;102:37–51 doi:10.1002/bdrc.21058.
55. Wit JM, Camacho-Hübner C. Endocrine regulation of longitudinal bone growth. Endocr Dev. 2011;21:30–41. doi:10.1159/000328119.
56. Yao Y, Wang Y. ATDC5: An excellent in vitro model cell line for skeletal development. J Cell Biochem. 2013;114:1223–9. doi:10.1002/jcb.24467.
57. Zhai Z, Yao Y, Wang Y. Importance of suitable reference Gene selection for quantitative RT-PCR during ATDC5 cells chondrocyte differentiation. PLoS One. 2013;8:e64786. doi:10.1371/journal.pone.0064786.

Pathogenetic Concepts of Joint Diseases

Jan Leipe and Hendrik Schulze-Koops

11.1	Joint Affection in Systemic Autoimmune Disorders – 174
11.2	Inflammatory Joint Diseases – 174
11.2.1	Synovial Inflammation – 175
11.2.2	Key Players in Synovial Inflammation – 175
11.3	Biologic Treatment Options in Rheumatoid Arthritis – 181
11.3.1	TNF Blockade – 182
11.3.2	IL-1 Neutralization – 182
11.3.3	IL-6 Blockade – 182
11.3.4	Co-stimulation Modulation – 183
11.3.5	B Cell Depletion – 183
	References – 184

What You Will Learn in This Chapter
In this book chapter, first you will learn how joints can be affected in different diseases with the focus on inflammatory joint disease. Second, current concepts of the pathogenesis of arthritis will be described and the key players of inflammation and joint destruction are discussed in detail. Third, established as well as new treatment options and how they target key players of joint disease are explained.

11.1 Joint Affection in Systemic Autoimmune Disorders

The wide spectrum of joint diseases in autoimmune disorders ranges from destructive/erosive processes affecting the synovium, cartilage, and bone, e.g., in rheumatoid arthritis (joint erosions), to ossifying changes with formation of new bone, e.g., in spondyloarthritis (syndesmophytes) or in osteoarthritis (osteophytes). Of note, both types of joint pathologies can be observed together in a certain disease, e.g., in psoriatic arthritis. These changes can be detected with good sensitivity and specificity by imaging studies including X-ray, ultrasound, magnetic resonance imaging, and computed tomography. On a molecular level, catabolic and anabolic molecular pathways are underlying the different patterns of joint diseases (destructive, proliferative). A dysfunction in joint remodeling results in an imbalance of degradation and formation of bone and cartilage. Whereas catabolic pathways such as those induced upon RANKL (receptor activator of nuclear factor κB (NF-κB) ligand)/ RANK, cathepsin K, and Dickkopf-1 (Dkk-1) engagement induce bone resorption and thereby erosive disease, anabolic pathways such as those involving bone morphogenic proteins (BMP) and wingless-type-like (Wnt) seem to favor new bone formation including osteophytes, syndesmophytes, and ankylosis. In addition, there are other pathways such as those mediated by hedgehog proteins that may have a dual function in arthritis, which are associated with catabolic or anabolic joint remodeling dependent on other factors. Future therapies might target these molecular pathways to specifically interfere with the imbalanced catabolic or anabolic joint remodeling in arthritis [1].

But not only cartilage and synovium – as central parts of the joint – can be affected in autoimmune diseases but also the fibrous parts such as the joint capsule and ligaments, e.g., in Jaccoud's arthropathy seen in connective tissue diseases, particularly systemic lupus erythematosus. Jaccoud's arthropathy mainly affecting the hands presents with «reversible» joint deformities secondary to soft tissue abnormalities (laxity of ligaments and joint capsule). The pathogenesis is poorly understood; synovitis is well documented and may somehow contribute to the process; however, it is not as aggressive and does not lead to erosive destruction [2].

Further, in psoriatic arthritis and other forms of spondyloarthritis, joints are often affected by periarticular inflammation contributing to joint pathology including tenosynovitis (inflammation of tendon sheaths), enthesitis (inflammation of the entheses, the sites where tendons or ligaments insert into the bone), and dactylitis (diffuse swelling of an entire finger or toe with a mixed picture of tenosynovitis and peritendinous soft tissue edema, enthesitis, and less frequently synovitis) [3].

11.2 Inflammatory Joint Diseases

Many different processes are involved in joint diseases. However, inflammation is involved in most aspects, not only in systemic autoimmune joint disorders but also in activated states of rather degenerative conditions such as osteoarthritis. Autoimmune inflammation is mostly systemic and can potentially cause pathologies in almost all organs but often affects the skin, the gastrointestinal tract; the cardiovascular, pulmonary, and central nervous system; and particularly the joints. In line with this, in rheumatoid arthritis (RA), although clinically presecting of a disease primarily of the joints, abnormal immune responses typically cause a variety of extra-articular manifestations. One of the mysteries of rheumatoid inflammation is why the synovium is the primary target.

11.2.1 Synovial Inflammation

The hallmark of joint inflammation is synovitis, which occurs when mononuclear cells (including T cells, B cells, plasma cells, dendritic cells, macrophages, and mast cells) infiltrate the synovial compartment or are locally activated or both. These infiltrations reflect more migration than local proliferation. Transendothelial leukocyte migration involves endothelial activation in synovial microvessels, which increases the expression of adhesion molecules such as integrins, selectins, and chemokines. Characteristic features of synovitis include neovascularization (induced by local hypoxic conditions and cytokines) promoting more inflammatory cell ingress into the synovium and insufficient lymphangiogenesis limiting the cellular egress. These microenvironmental changes, together with reorganization of the synovial architecture and local fibroblast activation, result in inflammatory synovial tissue observed particularly in RA [4]. The synovial lining then becomes hyperplastic, and the synovial membrane expands and forms villi. Enzymes secreted by neutrophils, synoviocytes, and chondrocytes degrade the cartilage, whereas the osteoclast-rich portion of the synovial membrane, or pannus, destroys the bone.

11.2.2 Key Players in Synovial Inflammation

The key players of synovial inflammation will be described based on the prototype and best-studied autoimmune joint disease, rheumatoid arthritis. The picture of pain, stiffness, swelling, and joint destruction seen in rheumatoid arthritis is a consequence of synovial inflammation, which is characterized by interactions of fibroblast-like synoviocytes with cells of the innate immune system, including macrophages, dendritic cells, mast cells, and NK cells, as well as cells of the adaptive immune system, B and T cells. In rheumatoid inflammation this process does not undergo the usual self-limitation characteristic of normal inflammatory reactions, rather chronicity ensues.

Innate Immune System in Rheumatoid Inflammation

Activation of innate immunity is probably the earliest process in rheumatoid inflammation. Innate immune cells such as macrophages, natural killer cells, and mast cells reside in the synovial membrane, while neutrophils are mainly found in the synovial fluid. Macrophages are matured and recruited to the synovium by factors including granulocyte-macrophage colony-stimulating factor (M-CSF, GM-CSF) and act as major effectors in synovitis by the release of pro-inflammatory cytokines, e.g., TNF or interleukin (IL)-1. Macrophages themselves are activated by (T cell-derived) cytokines, immune complexes, and a range of molecular pattern associated with pathogen and danger-associated molecular patterns (so-called PAMPs and DAMPs). Unlike adaptive immunity, innate immunity plays a role in a nonspecific recognition against microbial pathogens, e.g., bacteria or viruses. This primitive pattern-recognition system works through engagement of Toll-like receptors (TLRs), NOD-like receptors (NLRs), and RIG-like receptors (RLRs) by «nonself» microbial antigens, ultimately causing rapid inflammatory responses and clearance of the pathogen (PAMP). Whereas upon clearance of the pathogen, the inflammatory response is terminated, in rheumatoid inflammation, innate immune cells are persistently activated. In rheumatoid inflammation endogenous TLR ligands, such as HSP22, tenascin-C, serum amyloid A, and fragments of hyaluronic acid are upregulated leading to activation of monocytic cells through TLR 2 or 4 signaling [5]. Activation of the innate immune system also occurs in the noninfectious inflammatory response via so-called DAMPs, which are intracellular molecules whose release signals danger (e.g., HMGB, S100 proteins, ATP, and DNA). Expression of these molecules is observed, when cells are under stress, such as in chronic inflammation. In gouty arthritis, the activation of the innate immune system by DAMP constitutes a main pathogenic feature [6]. Here, the release of intracellular stores of uric acid, which belongs to DAMPs, and the consequent formation of monosodium urate crystals stimulate tissue-resident macrophages to produce pro-inflammatory mediators particularly IL-1 driving a robust inflammatory response.Despite the abundance of neutrophils in synovial fluid of rheumatoid inflammation, only a few infiltrate the synovium. Neutrophils contribute to synovitis by releasing prostaglandins, reactive oxygen and nitrogen species, and proteases into the extracellular space [7]. In this regard, GM-CSF, which is increased in RA synovial fluid, plasma, and synoviocytes, may play a role in the pathogenesis of RA through the activation, differentiation, and survival of neutrophils, macrophages, and eosinophils [8]. Neutralization of GM-CSF-mediated effects induced rapid clinically significant responses in RA patients in early studies, suggesting that inhibiting the mononuclear phagocyte pathway may provide a promising therapeutic approach for RA [9]. Mast cells are enriched in the RA synovial membrane, and their numbers correlate with lymphocytic infiltration and severity of synovitis [10]. They produce high levels of vasoactive amines, prostaglandin E2, cytokines (e.g., TNF), chemokines, and proteases (collagenase, tryptase). Finally, a subset of NK cells highly expressing CD56 is enriched in the synovium and favors maturation of dendritic cells and production of cytokines such as IFNγ and TNF, thereby promoting differentiation of pro-inflammatory T helper (Th) 1 cells [11]. However, NK cells in inflammatory conditions can also inhibit T cell expansion by mechanisms including lysis of effector T cells by perforin, produced mainly by NK cells with low expression of CD56 [12].

(Auto-)Antigen Presentation

Antigen-presenting cells (APCs) from the innate immune system and dendritic cells are loaded with disease-eliciting antigens most probably in the joint and then migrate to central lymphoid organs. Once there, APCs present an array of potential autoantigens to T cells, which can then stimulate B cells and migrate to the synovium. From what is known so far, it seems unlikely that a single autoantigen exists in RA. Instead, a broad spectrum of joint-specific antigens, such as type II collagen, or not joint-specific citrullinated antigens are involved. Citrullinated peptides including fibrinogen, vimentin, enolase, and collagen elicit immune responses more efficiently than the unmodified proteins. Of note, the initiating antigens may vary from patient to patient. Although innate and adaptive immune processes are involved in the initiation of arthritis, the genetics and the presence of autoantibodies clearly place adaptive immunity at the center of the pathogenesis of RA.

B Cells in Rheumatoid Inflammation

The precise role of B cells in rheumatoid inflammation is not fully understood yet; however, B cells contribute to joint pathogenesis by several mechanisms including autoantibody production, antigen presentation/T cell activation, and cytokine production.

The autoantibody repertoire in RA recognizing a large number of different antigens as stated above continues to expand. The two most widely studied autoantibody systems, whose importance is also emphasized by their inclusion in diagnosis and projection of the clinical prognosis of patients with RA, are rheumatoid factor and anti-citrullinated protein antibodies (ACPAs). The identification of rheumatoid factor as an autoantibody against the Fc portion of IgG was the first direct evidence that autoimmunity and B cells may play a role in RA. Rheumatoid factor and ACPAs can be found in most RA patients and are of importance for the diagnosis of RA, which is underlined by their inclusion in the new ACR/EULAR classification criteria for RA. Further, they are of major relevance for the prognosis since patients with rheumatoid factor and/or ACPA have more joint destruction, more extra-articular manifestations, and worse function [13]. However, it is speculated that these antibodies are not only disease markers but can also participate in disease process, e.g., by contributing to immune complex formation and complement activation. It has been shown that rheumatoid factor associated with immune complexes can enhance local inflammatory processes. In fact, early experiments have shown that injection of rheumatoid factor purified from a patient into a joint of the respective patient induced a marked inflammatory response compared to IgG control [13]. However, infusion of rheumatoid factor into healthy individuals caused neither sustained nor transient synovitis. This suggests that they exert pathogenic functions only in the context of rheumatoid inflammation, which is further supported by the notion that a considerable percentage of healthy individuals that never develop RA are seropositive.

In addition to being the precursors of antibody-secreting plasma cells, B cells act as highly efficient APCs to prime memory CD4 T cell differentiation and to activate T cells by co-stimulatory molecules (thereby also supporting autoreactive T cells). In fact, compared with nonspecific uptake associated with professional APCs, antigen-specific B cells can take up, process, and present peptides from nominal antigen with 1000-fold or greater efficiency. In addition, rheumatoid factor-positive B cells that are particularly believed to play a pathogenic role can take up antigen-Ig immune complexes via their membrane Ig receptors, which have rheumatoid factor specificity [13]. Besides antigen presentation, B cells potentially contribute to pathogenic T cell differentiation by secretion of cytokines such IL-1alpha, TNF, and IL-6, the latter (in the absence of TGF-β) inducing pro-inflammatory Th17 cells and diminishing regulatory T cells (Tregs) [14]. Further, B cells produce RANKL suggesting a direct involvement in osteoclastogenesis leading to bone resorption [14]. The role of B cells is supported by the therapeutic effect of B cell-targeted biologic agents. The success of B cell depletion therapy in RA may depend on disruption of all of these diverse functions.

T Cells in Rheumatoid Inflammation

The adaptive immune response is orchestrated by CD4 T cells. Full activation of T cells requires two signals. A first signal, which is antigen-specific, that recognizes its specific antigen presented in the context of two class II MHC molecules on the surface of APCs. A second signal, the co-stimulatory signal, is antigen nonspecific and is provided by the interaction between the co-stimulatory molecules CD80 and CD86 expressed on the membrane of APCs and the CD28 receptor expressed on the T cells. As a result of full activation, IL-2, the main growth factor for T cells, is secreted, and the IL-2 receptor is expressed. Additional signals, mainly cytokines, direct further Th differentiation into Th1, Th2, or Th17 effector cell subsets. Activation of naive T cells through the TCR in the absence of co-stimulation may lead to T cell anergy, T cell deletion, or the development of regulatory T cells (Tregs) [15].

Although the present understanding of the immunopathogenesis of rheumatoid inflammation is still incomplete, there is substantial evidence supporting the conclusion that CD4 T cells play a central role in both the initiation and the regulation of the chronic autoimmune response characteristic of the disease [16, 17]. The most compelling arguments for a direct contribution of CD4 T cells to the pathogenesis are the infiltration of the rheumatoid synovium with activated memory T cells, the presence of autoreactive T cells, the association of aggressive forms of the disease with particular HLA-DR alleles (whose only function is antigen presentation to T cells), and the recent clinical success of an inhibitor of T cell activation. Importantly, systemic inflammation in RA is characterized by a predominance of specialized inflammatory effector T cells, whereas T cell subsets with a potential to downmodulate inflammation, so-called Tregs are impaired. The limited efficacy of direct T cell targeting by cyclosporine or T cell depleting therapies is probably due to a broad and unselective deletion of regulatory as well as effector T cells indicating the need for targeting particular T cell subsets. Defining the contribution of T cells in the initiation and maintenance of inflammation has been augmented by the identification of functionally distinct subsets of effector Th cells that can be classified based on their cytokine and transcription factor profiles. In the recent years, in addition to the established Th cell subsets Th1 and Th2 cells, other subsets, such as Th17 and Th22 cells, have been described. Analysis of the role these T cell populations play in chronic inflammation provided insights into the pathogenic mechanisms of rheumatoid inflammation.

Th1 Cells in Rheumatoid Inflammation

Th1 cells are distinguished from other CD4 T cell subsets by expression of the transcription factor, T-bet, which mediates Th1 cell specific gene expression. They originate from naive T cells that were primed by IL-12. Th1 cells are characterized by production of their signature cytokine IFNγ and other pro-inflammatory cytokines including IL-2, lymphotoxin-α, and TNF. IFNγ induces a cellular immune response, in the physiological setting against intracellular bacteria, by induction of MHC class I and II expression, enhancement of antigen-processing pathways, and activation of cytotoxic CD8 T cells as well as NK cells [16]. Many studies in the 1990s suggested that rheumatoid inflammation is mediated by a predominance of activated Th1 cells over Th2 cells, the latter of which can downregulate inflammation. First studies demonstrated that the majority of T cell clones from the human rheumatoid synovial membrane functionally represent Th1 cells [18]. Consistent with this, large amounts of IFNγ were detectable in synovial biopsies, whereas IL-4 was rarely found [19, 20]. When migration of activated T cells into the inflamed synovium was blocked in patients with RA with a monoclonal antibody to ICAM-1, Th1 cells were trapped in the peripheral circulation, and this was associated with amelioration of disease activity [21]. Likewise, clinical efficacy after treatment with a monoclonal antibody to CD4 was associated with a dramatic decrease of Th1 cells, also suggesting that activated Th1 cells contribute to rheumatoid inflammation [22]. Finally, the analysis of synovial fluid showed enrichment of IFNγ-producing T cells, resulting in an elevated Th1/Th2 ratio, which correlated with disease activity [23, 24]. Clinical observations showing that the pregnancy-induced shift from a Th1 to a Th2 immune response seems to contribute to the gestational amelioration of RA further support a pathogenic role of Th1 cells in RA [16]. Another study demonstrated higher frequencies of citrulline-specific T cells (which have been implicated in the progression of experimental arthritis) in RA displaying a Th1 phenotype [25]. Anti-citrullinated protein antibodies are disease specific, suggesting an important etiologic role for immune recognition of self-proteins modified by citrullination.

Th17 Cells in Rheumatoid Inflammation

However, studies, including those with experimental models of arthritis demonstrating a paradoxical role for Th1 cell-derived IFNγ (in which the absence of an IFNγ response in genetically susceptible mice enhances autoimmune arthritis), gradually established that the Th1 phenotype failed to explain all the mechanisms involved in RA. Therefore, characterization of autoimmune responses as strictly Th1 or Th2 was thought to be overly simplistic [26, 27]. The identification of Th17 cells as a distinct subset of effector CD4 T cells has challenged the concept of a Th1/Th2 dichotomy, and it became unclear whether RA is a Th1- and/or a Th17 cell-mediated disease. Human Th17 cells are enforced by the transcription factor RORC (RORγT in mice) and produce IL-17 as their signature cytokine but not IL-4 or IFNγ [28–31]. Differentiation of these cells is promoted by a combination of cytokines including IL-6 (which suppresses the development of naive T cells into Tregs), IL-21, IL-23, and TGF-β. Th17 cells are further characterized by expression of IL-17F, IL-22, IL-23 receptor, and the chemokine receptor CCR6, the latter being of particular importance, because its ligand, CCL20, is expressed at high levels in inflamed joints, providing a basis for directed migration of Th17 cells into the inflamed synovium [32–34]. Consistent with this, IL-17 is highly expressed in RA synovium [35, 36] and in the synovial fluid of patients with early RA [37]. Moreover, high expression of IL-17 in RA synovial biopsies was associated with an increase in both activity and severity of the disease [38]. Consistent with this, cellular analyses in early but also in established RA revealed significantly increased Th17 cell frequencies and effector functions correlating with disease activity [39, 40]. In vivo, Th17 cells have been implicated as the pivotal driving force of autoimmune inflammation in several animal models of autoimmune arthritis, including collagen-induced arthritis (CIA) [41], and rat adjuvant-induced arthritis (AIA) [42]. The inhibition of Th17 cell development at the onset of the disease via the neutralization of IL-6, a critical Th17 cell-inducing cytokine in mice, suppresses the development of CIA [43], and neutralization of endogenous IL-17 during the initial phase of AIA prevents the onset of AIA [42]. IL-17 exerts joint-destructive activities. It contributes to synovitis by stimulating the production of the pro-inflammatory cytokines IL-1 and TNF from human macrophages and IL-6 from synovial fibroblasts; induces expression of RANKL, an important positive regulator of osteoclastogenesis; and also promotes cartilage degradation by inducing metalloproteinases and proteoglycan chondrocytes [40]. Studies blocking IL-17 showed efficacy in the treatment of RA and good efficacy in the treatment of psoriasis arthritis and ankylosing spondylitis [44–46].

Th22 Cells in Rheumatoid Inflammation

Lately, the existence of a new T cell subset was postulated which produces IL-22 but neither IL-17 nor IFNγ [47, 48] (Th22 cells). These cells have weaker pro-inflammatory effects compared to Th17 cells in coculture with fibroblast-like synoviocytes

[49]. In early RA, Th22 cells are increased as wells as IL-22-producing Th17 and Th1 cells [50]. These results, together with studies demonstrating the positive correlation of increased Th22 cells with DAS28 and CRP, emphasize their potential pathogenic role in RA [51, 52].

Tregs in Rheumatoid Inflammation

Tregs constitute a CD4 T cell population capable of suppressing primarily CD4 and CD8 effector T cell functions but also other immune cells, e.g., B cells, NK cells, and APCs. Although the exact mechanisms by which Tregs exert suppression are yet to be fully elucidated, it is thought that cell contact (e.g., CTLA-4) mechanisms and production of regulatory cytokines including IL-10 and TGF-β are involved in regulatory functions. Tregs are characterized by expression of the transcription factor Foxp3 and high levels of CD25, CTLA-4, GITR, and GARP all of which are implicated in their regulatory capacity [15]. Of note, with the potential exeption of GARP none of the surface molecules are completely specific for Tregs since they are also expressed on activated T cells, making the identification of Tregs based on specific markers and studies in human diseases difficult.

Tregs are involved in the onset and development of autoimmune diseases since their depletion in mice results in spontaneous development of several autoimmune diseases like autoimmune arthritis [53]. It is thought that RA arises from breakdown of self-tolerance leading to an aberrant immune response to autoantigens. The chronic nature of RA suggests a failure in mechanisms that control and suppress immune responses. In RA, the frequency of Tregs in the peripheral circulation is controversial with some studies reporting normal numbers, some an increase and others a decrease. Further, counterintuitively Tregs seem to be even enriched in RA synovium [54]. However, most reports demonstrate an impairment of Treg function in RA [55, 56]. These findings are supported by the observation that Tregs fail to control the activity of effector T cells during active disease [57]. The attenuated suppressive capacity seems to be caused by the pro-inflammatory milieu characteristic for rheumatoid inflammation, in particular cytokines like TNF, IL-6, or IL-21. Recently, it has been suggested that chronically active effector T cells in RA fail to produce sufficient amounts of IL-2, the primary activation factor for immunocompetent Tregs, and thereby escape Treg-mediated suppression. Treatment of patients with TNF-blocking agents on the other hand restored Treg function, which was associated with increased FOXP3 phosphorylation [57, 58]. Tregs are particularly unable to suppress pathogenic Th17 cells, which are enriched in RA [40, 59]. Efficient treatment with a TNF blocker was associated with a favorable, increased Treg/Th17 ratio [60, 61]. In line with this, altering the balance of Th17/Treg cells improves disease in the CIA model [62]. Impaired Treg function is also associated with reduced expression of CTLA-4 and impaired signaling in RA patients [59, 63]. In addition, distinct genetic and epigenetic factors including miRNA expression and DNA methylation may contribute to the altered Treg phenotype in RA [63–65].

Cytokines and Mesenchymal Tissue Response in Arthritis

A key feature of rheumatoid inflammation is the imbalanced expression of certain cytokines. Pro-inflammatory cytokine production that arises from different synovial cell populations is central to the pathogenesis. They orchestrate synovial inflammation, cartilage degradation induced by matrix metalloproteinases, and bone erosions mediated by osteoclast activation. Cytokine expression differs from patient to patient, and patterns may also change over time. However, some cytokines play a fundamental role in early as well as established disease. In addition to primarily T cell-derived cytokines such as IL-17 and IL-22 described above, the cytokines TNF, IL-1, and IL-6 are key mediators of inflammation and cartilage and bone destruction in RA and therefore play a dominant role in the pathobiology of rheumatoid inflammation.

TNF is a pleiotropic cytokine primarily produced by macrophages that mediates its diverse activities via two distinct receptors, TNF receptor I and II, thereby activating separate transduction pathways. In RA, where production of TNF is amplified and dysregulated, this cytokine can be detected at high levels in serum and synovial fluid. It contributes to rheumatoid inflammation and joint destruction through various mechanisms including monocyte activation; release of further pro-inflammatory cytokines, chemokines, prostaglandins, and MMPs primarily by macrophages and synovial fibroblasts; inhibition of Tregs; induction of endothelial cell adhesion molecules; promotion of angiogenesis; and pain. Thereby, TNF similar to IL-1 induces bone and proteoglycan resorption by synovial cells and inhibits formation of new cartilage and bone [66, 67]. In animal models overexpression of TNF leads to an aggressive and destructive synovitis and TNF blockade is anti-inflammatory [68]. Whereas the effect of TNF blockers in several arthritis models on bone and cartilage destruction is less prominent than that of IL-1 blockers, TNF inhibition in RA efficiently decreases joint matrix destruction, which can be seen by less radiographic progression [69].

Synovial macrophages are the major source of IL-1β in the joint, and almost half of them express IL-1β [70]. Immunohistologic studies located abundant IL-1β expression in synovial lining macrophages (type A synoviocytes) adjacent to fibroblast-like synoviocytes (type B synoviocytes) and in sublining macrophages in proximity to blood vessels [70]. IL-1β in the lining subsequently can activate synovial fibroblast proliferation and secretion of IL-6, IL-8, and GM-CSF. Further, IL-1β induces numerous adhesion molecules on fibroblast-like synoviocytes and endothelial cells, including VCAM-1 and ICAM-1, and enhances bone resorption [71]. In the joint IL-1, production can be triggered by immunoglobulin Fc fragments, immune complexes, collagen fragments, and type IX collagen.

IL-6 is produced mainly by fibroblast-like synoviocytes, macrophages, and also by T cells [70]. Particular IL-6 can promote activation and neutrophil recruitment to the joint directly through the IL-6 receptor [70]. IL-6 activates other local leukocyte subsets and fibroblasts and promotes the development of an imbalance between Th17 and Treg cells and the production of autoantibodies, such as rheumatoid factor and ACPA [72]. Systemic effects include the promotion of acute-phase responses, anemia, and lipid metabolism dysregulation. In patients with RA, IL-6 serum concentrations highly correlate with levels of acute-phase reactants such as C-reactive protein, α1-antitrypsin, fibrinogen, and haptoglobin [73]. Further, high levels of IL-6 can be found in RA synovial fluid.

11.3 Biologic Treatment Options in Rheumatoid Arthritis

Based on the increasing understanding of the molecular processes involved in the pathogenesis of RA, biological agents have been developed, targeting various immune mediators. Treatments blocking TNF-, IL-1-, and IL-6-mediated effects, as well as depletion and inhibition of B and T cells, has been proven effective in RA.

11.3.1 TNF Blockade

TNF was the first cytokine that was validated as a therapeutic target for RA through a series of preclinical studies of the immunobiology of synovial tissue/fluid, serum from patients, preclinical animal models, and, finally, clinical studies in RA patients. TNF blockers are the largest group of biologicals with currently five agents being approved for RA treatment. Of these, three are monoclonal antibodies (infliximab, adalimumab, and golimumab), one is a soluble TNF receptor Fc fusion protein (etanercept, the first agent introduced in 1998), and one is a humanized PEGylated Fab fragment (certolizumab). These agents, usually administered in combination with MTX, have shown efficacy in reducing signs and symptoms of RA, inhibiting the progression of structural damage and improving quality of life in the majority of patients [74]. Accumulating evidence suggests that early aggressive therapy with TNF blockers is particularly effective and can in some patients even induce long-term remission and prevent chronic synovitis. The observation that the patients who are not adequately responding to TNF blockade are still likely to benefit from the switch to other biologicals suggests that multiple parallel immune processes contribute to the pathogenesis.

11.3.2 IL-1 Neutralization

IL-1 has been implicated in the pathogenesis of RA, and indeed its inhibition using an IL-1 receptor antagonist (anakinra) leads to significant improvement in symptoms, laboratory parameters, and radiographic progression. However the clinical efficacy compared to other biologicals is rather modest, and improvement generally was observed in less than half of the patients. The use of anakinra in RA patients, who had failed TNF blockers, or also in the combination with TNF blockers provided no further benefit [75]. Some researchers attributed the limited efficacy of anakinra in RA to its short half-life indicating a need for higher continuous concentrations. However, other IL-1 inhibiting therapies including canakinumab, a monoclonal antibody against IL-1β with improved pharmacokinetics, showed clinical responses in RA similar to anakinra [76]. Anakinra was demonstrated very effective in diseases with a well-defined role for IL-1, such as Still's disease and cryopyrin-associated periodic syndromes. Together, although IL-1 certainly plays a role in rheumatoid inflammation, neutralizing IL-1 seems to be less efficacious in RA than, e.g., blocking TNF.

11.3.3 IL-6 Blockade

The first approved agent targeting the activity of IL-6 was tocilizumab, a humanized antibody against the IL-6 receptor (IL-6R). It blocks IL-6-mediated effects by inhibiting the binding of IL-6 to both transmembrane and soluble IL-6 receptors. Meanwhile also other IL-6 inhibitors are tested in clinical trials including a fully human anti-IL-6 receptor antibody (sarilumab), an anti-IL-6 receptor nanobody (ALX-0061), and anti-IL-6 antibodies (sirukumab, ALD518, olokizumab, and MEDI5117). Tocilizumab has already demonstrated effectiveness in reducing disease activity and radiographic progression similar to TNF inhibitors in several large phase III trials. The onset of response is quite rapid and can occur as early as 2–4 weeks after treatment initiation. In contrast to other biologicals, monotherapy

Pathogenetic Concepts of Joint Diseases

with tocilizumab was almost as effective as the combination of tocilizumab with MTX. This points to the pivotal role of IL-6 in the pathogenesis of RA and makes tocilizumab a good candidate for monotherapy when conventional DMARD therapy is not possible [75]. Further, in a randomized, double-blind trial, monotherapy with tocilizumab was superior to monotherapy with adalimumab in RA [77]. Of note, due to the direct and pronounced effects of IL-6 blockade on production of acute-phase reactants in the liver, caution should be used when interpreting results for inflammation markers, particularly CRP values.

11.3.4 Co-stimulation Modulation

Abatacept is a soluble fusion protein that consists of human CTLA-4 linked to a modified Fc portion of human IgG1 (CTLA4-Ig). Abatacept is a co-stimulation modulator that inhibits T cell activation by preventing CD28 from binding to its counter receptor, CD80/CD86, due to its higher affinity for CD80/CD86, thereby blocking interaction between CD80/CD86 on APC and CD28 on T cells. This blockade of the co-stimulatory signal prevents the full activation of T cells provided when CD80/CD86 normally interact with CD28. In early studies administration of CTLA4-Ig or successful transfer of its coding nucleic acid prevented or ameliorated CIA [78]. Abatacept has demonstrated improvement of signs and symptoms of RA, physical function, and radiographic progression of structural damage in a wide range of early and long-standing RA populations. It was therefore approved for the treatment of patients with moderately to severely active RA who have had an inadequate response to one or more conventional DMARDs or TNF inhibitors [74]. Given its favorable safety profile, which even appears slightly better than most other biologicals, in the new ACR guidelines on RA, abatacept was recommended in case of serious infection over TNF inhibitors. The success of this treatment principle supports the important role for T cells in this disease pathobiology.

11.3.5 B Cell Depletion

B cell depletion using the anti-CD20 chimeric monoclonal antibody, rituximab, was demonstrated to be efficacious in RA patients. It has been approved for the treatment of moderate-to-severe RA, in combination with MTX, and in patients who have had an inadequate response to at least one TNF inhibitor. Rituximab treatment results in significant improvements after 8–16 weeks. However, in patients with only a partial response after 6 months, additional treatment courses often achieve a good response. Interestingly, patients who are positive for rheumatoid factor or ACPA show a better treatment response with rituximab than seronegative patients. B cell depletion is caused by several antibody-dependent mechanisms including Fc receptor gamma-mediated antibody-dependent cytotoxicity (regarded as the principle mechanism), complement-mediated cell lysis, growth arrest, and B cell apoptosis. Although CD20 is not expressed on long-lived antibody-producing plasma cells, the observed decrease of pathophysiological important autoantibodies after rituximab suggests that these autoantibodies are produced at least in part by short-lived plasma cells. An indirect mechanism of action seems to be a substantial decrease of CD4 T cells following rituximab [79]. The relevance of these findings is supported by the observation that a lack of CD4 T cell depletion was associated with an inadequate clinical response to rituximab [79].

Several meta-analyses have shown favorable and similar efficacy among the available biologicals (except anakinra). However, as many immune mediators work in concert during the disease process, it is still unclear what should be targeted first, cytokines, T cells, or B cells. Further, exact immunological mechanisms of biologicals including their net effects on different immune cells and cytokines/chemokines need to be investigated in order to understand the immune network. Finally, there are also clinical challenges, most notably safety issues that are associated with immunosuppression resulting in increased infections and a slightly increased incidence of tumors, particular lymphomas.

> **Take-Home Messages**
> - Joints can be affected in several ways in different diseases, e.g., erosive processes as well as ossifying changes with formation of new bone based on the balance of catabolic and anabolic molecular pathways.
> - Inflammatory joint disease is characterized by infiltration of mononuclear immune cells in the synovial and surrounding structures.
> - Immune cells, such as pro-inflammatory T cells, B cells, and macrophages as well as mesenchymal cells, e.g., fibroblasts, are key players in joint inflammation and mediate their joint destructive processes particularly by inflammatory cytokine/chemokines.
> - Modern therapies for inflammatory joint disease target key players of arthritis, e.g., abatacept – T cells, Rituximab – B cells, anti-TNF/IL-6/IL-17 agents – cytokines with great success.

References

1. Beyer C, Schett G. Pharmacotherapy: concepts of pathogenesis and emerging treatments. Novel targets in bone and cartilage. Best Pract Res Clin Rheumatol. 2010;24(4):489–96.
2. Santiago MB. Miscellaneous non-inflammatory musculoskeletal conditions. Jaccoud's arthropathy. Best Pract Res Clin Rheumatol. 2011;25(5):715–25.
3. Sakkas LI, Alexiou I, Simopoulou T, Vlychou M. Enthesitis in psoriatic arthritis. Semin Arthritis Rheum. 2013;43(3):325–34.
4. Szekanecz Z, Pakozdi A, Szentpetery A, Besenyei T, Koch AE. Chemokines and angiogenesis in rheumatoid arthritis. Front Biosci (Elite Ed). 2009;1:44–51.
5. Gierut A, Perlman H, Pope RM. Innate immunity and rheumatoid arthritis. Rheum Dis Clin N Am. 2010;36(2):271–96.
6. Rock KL, Kataoka H, Lai JJ. Uric acid as a danger signal in gout and its comorbidities. Nat Rev Rheumatol. 2013;9(1):13–23.
7. Cascao R, Rosario HS, Souto-Carneiro MM, Fonseca JE. Neutrophils in rheumatoid arthritis: more than simple final effectors. Autoimmun Rev. 2010;9(8):531–5.
8. Greven DE, Cohen ES, Gerlag DM, Campbell J, Woods J, Davis N, et al. Preclinical characterisation of the GM-CSF receptor as a therapeutic target in rheumatoid arthritis. Ann Rheum Dis. 2014;74(10): 1924–30.
9. Burmester GR, Weinblatt ME, McInnes IB, Porter D, Barbarash O, Vatutin M, et al. Efficacy and safety of mavrilimumab in subjects with rheumatoid arthritis. Ann Rheum Dis. 2013;72(9):1445–52.
10. Malone DG, Wilder RL, Saavedra-Delgado AM, Metcalfe DD. Mast cell numbers in rheumatoid synovial tissues. Correlations with quantitative measures of lymphocytic infiltration and modulation by antiinflammatory therapy. Arthritis Rheum. 1987;30(2):130–7.
11. Daien CI, Gailhac S, Audo R, Mura T, Hahne M, Combe B, et al. High levels of natural killer cells are associated with response to tocilizumab in patients with severe rheumatoid arthritis. Rheumatology (Oxford). 2014;54(4):601–8.

12. Fort MM, Leach MW, Rennick DM. A role for NK cells as regulators of CD4+ T cells in a transfer model of colitis. J Immunol. 1998;161(7):3256–61.
13. Silverman GJ, Carson DA. Roles of B cells in rheumatoid arthritis. Arthritis Res Ther. 2003;5(Suppl 4):S1–6.
14. Bugatti S, Vitolo B, Caporali R, Montecucco C, Manzo A. B cells in rheumatoid arthritis: from pathogenic players to disease biomarkers. Biomed Res Int. 2014;2014:681678.
15. Leipe J, Skapenko A, Lipsky PE, Schulze-Koops H. Regulatory T cells in rheumatoid arthritis. Arthritis Res Ther. 2005;7(3):93.
16. Skapenko A, Leipe J, Lipsky PE, Schulze-Koops H. The role of the T cell in autoimmune inflammation. Arthritis Res Ther. 2005;7(Suppl 2):S4–14.
17. Leipe J, Chang HD. Effector T cells. Z Rheumatol. 2015;74(1):14–9.
18. Miltenburg AM, van Laar JM, de Kuiper R, Daha MR, Breedveld FC. T cells cloned from human rheumatoid synovial membrane functionally represent the Th1 subset. Scand J Immunol. 1992;35:603–10.
19. Kusaba M, Honda J, Fukuda T, Oizumi K. Analysis of type 1 and type 2 T cells in synovial fluid and peripheral blood of patients with rheumatoid arthritis. J Rheumatol. 1998;25:1466–71.
20. Canete JD, Martinez SE, Farres J, Sanmarti R, Blay M, Gomez A, et al. Differential Th1/Th2 cytokine patterns in chronic arthritis: interferon gamma is highly expressed in synovium of rheumatoid arthritis compared with seronegative spondyloarthropathies. Ann Rheum Dis. 2000;59(4):263–8.
21. Schulze-Koops H, Lipsky PE, Kavanaugh AF, Davis LS. Elevated Th1- or Th0-like cytokine mRNA in peripheral circulation of patients with rheumatoid arthritis: modulation by treatment with anti-ICAM-1 correlates with clinical benefit. J Immunol. 1995;155:5029–37.
22. Schulze-Koops H, Davis LS, Haverty TP, Wacholtz MC, Lipsky PE. Reduction of Th1 cell activity in the peripheral circulation of patients with rheumatoid arthritis after treatment with a non-depleting humanized monoclonal antibody to CD4. J Rheumatol. 1998;25:2065–76.
23. Davis LS, Schulze-Koops H, Lipsky PE. Rheumatoid synovial memory T cells have been primed in vivo to be interferon-gamma producers. Arthritis Rheum. 1997;40:S532.
24. Miyata M, Ohira H, Sasajima T, Suzuki S, Ito M, Sato Y, et al. Significance of low mRNA levels of interleukin-4 and -10 in mononuclear cells of the synovial fluid of patients with rheumatoid arthritis. Clin Rheumatol. 2000;19(5):365–70.
25. James EA, Rieck M, Pieper J, Gebe JA, Yue BB, Tatum M, et al. Citrulline-specific Th1 cells are increased in rheumatoid arthritis and their frequency is influenced by disease duration and therapy. Arthritis Rheumatol. 2014 Jul;66(7):1712–22.
26. Vermeire K, Heremans H, Vandeputte M, Huang S, Billiau A, Matthys P. Accelerated collagen-induced arthritis in IFN-gamma receptor-deficient mice. J Immunol. 1997;158(11):5507–13.
27. Manoury-Schwartz B, Chiocchia G, Bessis N, Abehsira-Amar O, Batteux F, Muller S, et al. High susceptibility to collagen-induced arthritis in mice lacking IFN-gamma receptors. J Immunol. 1997;158(11):5501–6.
28. Ivanov II, McKenzie BS, Zhou L, Tadokoro CE, Lepelley A, Lafaille JJ, et al. The orphan nuclear receptor RORgammat directs the differentiation program of proinflammatory IL-17+ T helper cells. Cell. 2006;126(6):1121–33.
29. Manel N, Unutmaz D, Littman DR. The differentiation of human T(H)-17 cells requires transforming growth factor-beta and induction of the nuclear receptor RORgammat. Nat Immunol. 2008;9(6):641–9.
30. Harrington LE, Hatton RD, Mangan PR, Turner H, Murphy TL, Murphy KM, et al. Interleukin 17-producing CD4+ effector T cells develop via a lineage distinct from the T helper type 1 and 2 lineages. Nat Immunol. 2005;6(11):1123–32.
31. Park H, Li Z, Yang XO, Chang SH, Nurieva R, Wang YH, et al. A distinct lineage of CD4 T cells regulates tissue inflammation by producing interleukin 17. Nat Immunol. 2005;6(11):1133–41.
32. Acosta-Rodriguez EV, Rivino L, Geginat J, Jarrossay D, Gattorno M, Lanzavecchia A, et al. Surface phenotype and antigenic specificity of human interleukin 17-producing T helper memory cells. Nat Immunol. 2007;8(6):639–46.
33. Hirota K, Yoshitomi H, Hashimoto M, Maeda S, Teradaira S, Sugimoto N, et al. Preferential recruitment of CCR6-expressing Th17 cells to inflamed joints via CCL20 in rheumatoid arthritis and its animal model. J Exp Med. 2007;204(12):2803–12.
34. Lubberts E. Role of T lymphocytes in the development of rheumatoid arthritis. Implications for treatment. Curr Pharm Des. 2015;21(2):142–6.
35. Chabaud M, Durand JM, Buchs N, Fossiez F, Page G, Frappart L, et al. Human interleukin-17: a T cell-derived proinflammatory cytokine produced by the rheumatoid synovium. Arthritis Rheum. 1999;42(5):963–70.

36. Kotake S, Udagawa N, Takahashi N, Matsuzaki K, Itoh K, Ishiyama S, et al. IL-17 in synovial fluids from patients with rheumatoid arthritis is a potent stimulator of osteoclastogenesis. J Clin Invest. 1999;103(9):1345–52.
37. Raza K, Falciani F, Curnow SJ, Ross EJ, Lee CY, Akbar AN, et al. Early rheumatoid arthritis is characterized by a distinct and transient synovial fluid cytokine profile of T cell and stromal cell origin. Arthritis Res Ther. 2005;7(4):R784–95.
38. Kirkham BW, Lassere MN, Edmonds JP, Juhasz KM, Bird PA, Lee CS, et al. Synovial membrane cytokine expression is predictive of joint damage progression in rheumatoid arthritis: a two-year prospective study (the DAMAGE study cohort). Arthritis Rheum. 2006;54(4):1122–31.
39. Shen H, Goodall JC, Hill Gaston JS. Frequency and phenotype of peripheral blood Th17 cells in ankylosing spondylitis and rheumatoid arthritis. Arthritis Rheum. 2009;60(6):1647–56.
40. Leipe J, Grunke M, Dechant C, Reindl C, Kerzendorf U, Schulze-Koops H, et al. Role of Th17 cells in human autoimmune arthritis. Arthritis Rheum. 2010;62(10):2876–85.
41. Nakae S, Nambu A, Sudo K, Iwakura Y. Suppression of immune induction of collagen-induced arthritis in IL-17-deficient mice. J Immunol. 2003;171(11):6173–7.
42. Bush KA, Farmer KM, Walker JS, Kirkham BW. Reduction of joint inflammation and bone erosion in rat adjuvant arthritis by treatment with interleukin-17 receptor IgG1 Fc fusion protein. Arthritis Rheum. 2002;46(3):802–5.
43. Miossec P, Korn T, Kuchroo VK. Interleukin-17 and type 17 helper T cells. N Engl J Med. 2009;361(9):888–98.
44. Genovese MC, Durez P, Richards HB, Supronik J, Dokoupilova E, Aelion JA, et al. One-year efficacy and safety results of secukinumab in patients with rheumatoid arthritis: phase II, dose-finding, double-blind, randomized, placebo-controlled study. J Rheumatol. 2014;41(3):414–21.
45. Mease PJ, McInnes IB, Kirkham B, Kavanaugh A, Rahman P, van der Heijde D, et al. Secukinumab inhibition of interleukin-17A in patients with psoriatic arthritis. N Engl J Med. 2015;373(14):1329–39.
46. Baeten D, Sieper J, Braun J, Baraliakos X, Dougados M, Emery P, et al. Secukinumab, an interleukin-17A inhibitor, in Ankylosing spondylitis. N Engl J Med. 2015;373(26):2534–48.
47. Duhen T, Geiger R, Jarrossay D, Lanzavecchia A, Sallusto F. Production of interleukin 22 but not interleukin 17 by a subset of human skin-homing memory T cells. Nat Immunol. 2009;10(8):857–63.
48. Trifari S, Kaplan CD, Tran EH, Crellin NK, Spits H. Identification of a human helper T cell population that has abundant production of interleukin 22 and is distinct from T(H)-17, T(H)1 and T(H)2 cells. Nat Immunol. 2009;10(8):864–71.
49. van Hamburg JP, Corneth OB, Paulissen SM, Davelaar N, Asmawidjaja PS, Mus AM, et al. IL-17/Th17 mediated synovial inflammation is IL-22 independent. Ann Rheum Dis. 2013;72(10):1700–7.
50. Leipe J, Schramm MA, Grunke M, Baeuerle M, Dechant C, Nigg AP, et al. Interleukin 22 serum levels are associated with radiographic progression in rheumatoid arthritis. Ann Rheum Dis. 2011;70(8):1453–7.
51. Ren J, Feng Z, Lv Z, Chen X, Li J. Natural killer-22 cells in the synovial fluid of patients with rheumatoid arthritis are an innate source of interleukin 22 and tumor necrosis factor-{alpha}. J Rheumatol. 2011;38(10):2112–8.
52. Zhang L, Li JM, Liu XG, Ma DX, Hu NW, Li YG, et al. Elevated Th22 cells correlated with Th17 cells in patients with rheumatoid arthritis. J Clin Immunol. 2011;31(4):606–14.
53. Sakaguchi S, Sakaguchi N, Asano M, Itoh M, Toda M. Immunologic self-tolerance maintained by activated T cells expressing IL-2 receptor alpha-chains (CD25). Breakdown of a single mechanism of self-tolerance causes various autoimmune diseases. J Immunol. 1995;155(3):1151–64.
54. Jiao Z, Wang W, Jia R, Li J, You H, Chen L, et al. Accumulation of FoxP3-expressing CD4+CD25+ T cells with distinct chemokine receptors in synovial fluid of patients with active rheumatoid arthritis. Scand J Rheumatol. 2007;36(6):428–33.
55. Beavis PA, Gregory B, Green P, Cribbs AP, Kennedy A, Amjadi P, et al. Resistance to regulatory T cell-mediated suppression in rheumatoid arthritis can be bypassed by ectopic foxp3 expression in pathogenic synovial T cells. Proc Natl Acad Sci U S A. 2011;108(40):16717–22.
56. Cooles FA, Isaacs JD, Anderson AE. Treg cells in rheumatoid arthritis: an update. Curr Rheumatol Rep. 2013;15(9):352.
57. Ehrenstein MR, Evans JG, Singh A, Moore S, Warnes G, Isenberg DA, et al. Compromised function of regulatory T cells in rheumatoid arthritis and reversal by anti-TNFalpha therapy. J Exp Med. 2004;200(3):277–85.
58. Nie H, Zheng Y, Li R, Guo TB, He D, Fang L, et al. Phosphorylation of FOXP3 controls regulatory T cell function and is inhibited by TNF-alpha in rheumatoid arthritis. Nat Med. 2013;19(3):322–8.

59. Flores-Borja F, Jury EC, Mauri C, Ehrenstein MR. Defects in CTLA-4 are associated with abnormal regulatory T cell function in rheumatoid arthritis. Proc Natl Acad Sci U S A. 2008;105(49):19396–401.
60. McGovern JL, Nguyen DX, Notley CA, Mauri C, Isenberg DA, Ehrenstein MR. Th17 cells are restrained by Treg cells via the inhibition of interleukin-6 in patients with rheumatoid arthritis responding to anti-tumor necrosis factor antibody therapy. Arthritis Rheum. 2012;64(10):3129–38.
61. Huang Z, Yang B, Shi Y, Cai B, Li Y, Feng W, et al. Anti-TNF-alpha therapy improves Treg and suppresses Teff in patients with rheumatoid arthritis. Cell Immunol. 2012;279(1):25–9.
62. Park JS, Lim MA, Cho ML, Ryu JG, Moon YM, Jhun JY, et al. p53 controls autoimmune arthritis via STAT-mediated regulation of the Th17 cell/Treg cell balance in mice. Arthritis Rheum. 2013;65(4):949–59.
63. Cribbs AP, Kennedy A, Penn H, Read JE, Amjadi P, Green P, et al. Treg cell function in rheumatoid arthritis is compromised by ctla-4 promoter methylation resulting in a failure to activate the indoleamine 2,3-dioxygenase pathway. Arthritis Rheumatol. 2014;66(9):2344–54.
64. Zhou Q, Haupt S, Kreuzer JT, Hammitzsch A, Proft F, Neumann C, et al. Decreased expression of miR-146a and miR-155 contributes to an abnormal Treg phenotype in patients with rheumatoid arthritis. Ann Rheum Dis. 2014;74(6):1265–74.
65. Cribbs AP, Kennedy A, Penn H, Amjadi P, Green P, Read JE, et al. Methotrexate restores regulatory T cell function through demethylation of the foxp3 upstream enhancer in patients with rheumatoid arthritis. Arthritis Rheumatol. 2015;67(5):1182–92.
66. Saklatvala J. Tumour necrosis factor alpha stimulates resorption and inhibits synthesis of proteoglycan in cartilage. Nature. 1986;322(6079):547–9.
67. Bertolini DR, Nedwin GE, Bringman TS, Smith DD, Mundy GR. Stimulation of bone resorption and inhibition of bone formation in vitro by human tumour necrosis factors. Nature. 1986;319(6053):516–8.
68. Keffer J, Probert L, Cazlaris H, Georgopoulos S, Kaslaris E, Kioussis D, et al. Transgenic mice expressing human tumour necrosis factor: a predictive genetic model of arthritis. EMBO J. 1991;10:4025–31.
69. Lipsky PE, van der Heijde DM, St Clair EW, Furst DE, Breedveld FC, Kalden JR, et al. Infliximab and methotrexate in the treatment of rheumatoid arthritis. Anti-tumor necrosis factor trial in rheumatoid arthritis with concomitant therapy study group. N Engl J Med. 2000;343(22):1594–602.
70. Firestein GS, Alvaro-Gracia JM, Maki R. Quantitative analysis of cytokine gene expression in rheumatoid arthritis. [published erratum appears in J Immunol 1990 Aug 1; 145(3):1037]. JImmunol. 1990;144:3347–53.
71. McInnes IB, Schett G. Cytokines in the pathogenesis of rheumatoid arthritis. Nat Rev Immunol. 2007;7(6):429–42.
72. Leipe J, Skapenko A, Schulze-Koops H. Th17 cells - a new proinflammatory T cell population and its role in rheumatologic autoimmune diseases. Z Rheumatol. 2009;68(5):405–8.
73. Guerne PA, Zuraw BL, Vaughan JH, Carson DA, Lotz M. Synovium as a source of interleukin 6 in vitro. Contribution to local and systemic manifestations of arthritis. J Clin Invest. 1989;83(2):585–92.
74. Nam JL, Ramiro S, Gaujoux-Viala C, Takase K, Leon-Garcia M, Emery P, et al. Efficacy of biological disease-modifying antirheumatic drugs: a systematic literature review informing the 2013 update of the EULAR recommendations for the management of rheumatoid arthritis. Ann Rheum Dis. 2014;73(3):516–28.
75. Furst DE, Emery P. Rheumatoid arthritis pathophysiology: update on emerging cytokine and cytokine-associated cell targets. Rheumatology (Oxford). 2014;53(9):1560–9.
76. Alten R, Gomez-Reino J, Durez P, Beaulieu A, Sebba A, Krammer G, et al. Efficacy and safety of the human anti-IL-1beta monoclonal antibody canakinumab in rheumatoid arthritis: results of a 12-week, phase II, dose-finding study. BMC Musculoskelet Disord. 2011;12:153.
77. Gabay C, Emery P, van Vollenhoven R, Dikranian A, Alten R, Pavelka K, et al. Tocilizumab monotherapy versus adalimumab monotherapy for treatment of rheumatoid arthritis (ADACTA): a randomised, double-blind, controlled phase 4 trial. Lancet. 2013;381(9877):1541–50.
78. Quattrocchi E, Dallman MJ, Feldmann M. Adenovirus-mediated gene transfer of CTLA-4Ig fusion protein in the suppression of experimental autoimmune arthritis. Arthritis Rheum. 2000;43(8):1688–97.
79. Melet J, Mulleman D, Goupille P, Ribourtout B, Watier H, Thibault G. Rituximab-induced T cell depletion in patients with rheumatoid arthritis: association with clinical response. Arthritis Rheum. 2013;65(11):2783–90.

Osteoarthritis Biology

Stefan Toegel

12.1 **Osteoarthritis – 190**
12.1.1 Risk Factors – 190
12.1.2 Socioeconomic Impact – 191
12.1.3 Pathophysiology of OA – 191

12.2 **Triggers of OA Biology – 194**
12.2.1 Ageing – 194
12.2.2 Obesity – 195
12.2.3 Genetics – 195
12.2.4 Inflammation – 196
12.2.5 Subchondral Bone – 198
12.2.6 Signalling – 199

12.3 **Concluding Remarks – 201**

References – 202

© Springer International Publishing AG 2017
P. Pietschmann (ed.), *Principles of Bone and Joint Research*, Learning Materials in Biosciences,
DOI 10.1007/978-3-319-58955-8_12

What You Will Learn in This Chapter

In today's ageing society, osteoarthritis has become a major challenge for health-care systems and research alike. Although the molecular mechanisms of tissue degradation are well explored and functional disease markers have been defined, important questions on the pathophysiology of OA still remain unanswered. This chapter provides an overview of the socioeconomic relevance and risk factors of OA, highlights the key features of OA pathology including matrix degradation and pro-inflammatory signalling and discusses current concepts and future perspectives of OA research. It is envisioned that this brief primer to OA biology contributes to elucidate disease mechanisms and to foster critical discussion and concerted scientific efforts in the field of OA.

12.1 Osteoarthritis

Osteoarthritis (OA) is a degenerative disease of the joint, clinically characterized by pain, stiffness, joint effusion and loss of joint mobility. The erosion of cartilaginous tissues from articular surfaces within the joint and the inability of the tissue to effectively repair and regenerate these regions lead to irreversible loss of joint functionality. However, OA is far beyond a simple «wear and tear» of articular cartilage. In fact, OA affects the whole joint involving synovial inflammation, subchondral bone sclerosis and destruction of articular cartilage [1]. In addition, OA has significant cellular and molecular components that drive disease progression via pro-inflammatory and phenotype-modulating signalling cascades.

12.1.1 Risk Factors

OA is a multifactorial disease. The most consistent risk factors for OA are age, female gender, overweight/obesity and race/ethnicity. Although OA is clearly an ageing-related disease, age alone does not cause OA but rather promotes the development of OA together with other risk factors. The strong relationship between age, obesity, inflammation and OA, as well as its underlying molecular mechanisms, are discussed in dedicated sections below. Female sex is associated with higher risk and severity of OA. Although not verified yet, this association might stem from the observation that women have thinner cartilage with more reduced volumes as compared to men, which could result in a more accelerated loss of cartilage under OA conditions [2]. Studies on role of sex hormones such as oestrogen in OA have yielded conflicting results. Whether there is a causal relationship between the loss of oestrogen during menopause and the onset of OA or whether reducing oestrogen levels result in higher pain sensitivity that unmasks OA symptoms is under discussion.

Besides, all causes that result in abnormal loading of the joints such as malalignment or previous joint injuries are associated with OA. It has been shown that over 30% of patients with acute anterior cruciate ligament or meniscal injuries develop radiographic knee OA within 5 years post-injury, and over 50% of patients show evidence 10–20 years post-injury [3]. Evidence has accumulated that joint injuries induce biomechanical changes as well as biochemical cascades (e.g. inflammation) that can eventually lead to the development of OA.

12.1.2 Socioeconomic Impact

OA belongs to the most widespread diseases, particularly in the aged population. In 2005, it was estimated that over 26 million people in the USA had some form of OA [4]. The Rotterdam study showed that 15% and 6% of the more than thousand participants (age 68.2 ± 8.0 years) had radiographic signs of OA of the knee and hip, respectively [5]. Whereas symptomatic OA is always associated with radiologic signs, not all patients with radiographic OA have concomitant symptoms. Thus, estimates for symptomatic OA might be lower. The WHO estimated that 10–20% of the world's population aged 60 years or older have significant clinical problems attributed to OA [6].

Unfortunately, no effective disease-modifying OA drugs are currently available. OA characteristically remains asymptomatic for a long time and is frequently diagnosed very late in disease progression when irreversible structural damages have already occurred. This eventually leads to disability or necessitates total arthroplasty resulting in tremendous costs for health-care systems. For example, according to the Federal Statistical Office of Germany (Destatis), in 2008 the estimated direct costs (i.e. excluding costs owing to disability or early retirement) spent for OA treatments in Germany were about 8 billion euros [7]. In Europe, joint replacement surgeries to treat OA are performed every 1.5 minutes, whereas in the USA a total of approximately 500,000 joint replacements are performed each year [8]. OA has a negative impact on societies by contributing to work loss, disability pensions, early retirement and the increasing need for social support [9]. In Vienna, degenerative joint diseases are among the leading causes for sickness absence, rehabilitation and hospitalization [10].

Age is most probably the key factor of the increase in OA prevalence that has been recognized over the last decades. Due to demographic changes and the progressive impact of related risk factors such as obesity, OA prevalence is expected to further increase rapidly. According to the United Nations, the world population is predicted to reach 9.6 billion by 2050 [11]. People aged over 60 years will then account for more than 20% of the world's population, of which, according to a conservative estimate, 15% will have symptomatic OA, and one-third of these people will be severely disabled. Similarly, a recent study predicted that there is a risk for a 50% increase in OA prevalence in 20 years from now [12].

Consequently, OA represents a major socioeconomic issue that induces a heavy burden on health-care systems worldwide, which will even gain importance due to increased life expectancy.

12.1.3 Pathophysiology of OA

Key features of osteoarthritic joints include deformation of the articular surface, accompanied by chondrocyte proliferation and loss of matrix proteoglycans, subchondral bone thickening, osteophyte formation, synovial intima cell hyperplasia and synovial fibrosis [13–17].

Macroscopically, changes of OA cartilage can be described as softening, fibrillation and erosion. Histologic features of cartilage breakdown include cartilage clefts, loss of cartilage layers, cellular necrosis, chondrocyte cloning and a duplication of the tidemark. It appears that the superficial zone of cartilage is affected first in early OA. At the biochemical level, the OA process is directly linked to the loss of proteoglycan content and the altered composition of extracellular matrix (ECM) components. As such, proteoglycans of the non-aggregated form (unbound to hyaluronate) accumulate in OA cartilage. Furthermore, proteolytic degradation of proteoglycans reduces the chain length of the proteoglycans and inhibits the formation of normal macromolecular complexes, leading to a more permeable matrix and a significant diminution of the hydraulic pressure in early OA cartilage [13]. In addition, the three-dimensional (3D) collagen network is severely perturbed in OA, which can be visualized by polarized light microscopy or T2 relaxation time as a reproducible MRI parameter. In summary, the breakdown of ECM architecture in OA cartilage causes a reduction in the compressive stiffness and loss of tissue functionality.

Phenotypic Modulation of Chondrocytes

Articular cartilage is a unique and unusual tissue with a single resident cell type, the chondrocyte. Classically, chondrocyte phenotypes are categorized largely by subtyping of collagen gene expression: Chondroprogenitor cells are characterized by the expression of the alternative splice variant of type II collagen, type IIA procollagen (COL2A). Mature chondrocytes express the typical cartilage collagen types II (COL2B), IX and XI as well as aggrecan and link protein. Hypertrophic chondrocytes found in the lowest zone of the cartilage of the foetal growth plate band in the calcified zone of adult cartilage are marked by the expression of type X collagen [17].

Unlike the growth plate, which is a transient structure to be replaced by the bone, articular cartilage is designed to be permanent, ideally remaining functional for the entire lifetime of the individual. Whereas growth plate chondrocytes typically undergo rapid proliferation followed by cellcycle exit, hypertrophy and apoptosis, chondrocytes of articular cartilage are phenotypically stable and responsible for ECM synthesis, maintenance and degradation.

Under OA conditions, however, chondrocytes undergo a range of phenotypic modulations that represent characteristic features of the disease. Evidence from basic research indicates that a significant proportion of OA chondrocytes starts to re-express a chondroprogenitor phenotype, which is comparable to that observed in foetal skeletal development [18]. This switch from the articular to the growth plate chondrocyte phenotype is accompanied by the re-initiation of proliferation and hypertrophic differentiation [19]. In contrast to normal articular chondrocytes, which have essentially no proliferative activity, OA chondrocytes start to exhibit a very low proliferative activity that most probably is responsible for the chondrocyte clustering observed in OA cartilage. In OA cartilage lesions, markers of hypertrophic chondrocytes (type X collagen, osteocalcin, CD36 and alkaline phosphatase) were found to be up-regulated. Key features of chondrocyte hypertrophy – such as downregulation of SOX9, type II collagen and aggrecan expression, induction of matrix metalloproteinase 13 (MMP-13) expression, chondrocyte apoptosis, ECM mineralization and recruitment of blood vessels and osteoclasts – might have a direct input to joint destruction in OA. In addition, hypertrophic chondrocytes might drive cartilage loss by directing the behaviour of other cell types (endothelial cells, osteoblasts and osteoclasts) to replace cartilage with bone. It is noteworthy, however, that the concept of chondrocyte hypertrophy in OA is based on the evaluation of molecular characteristics (such as the expression of type X collagen) in OA cartilage. The essential feature of hypertrophic chondrocytes, namely, large cell volume, is hardly observed in hypertrophic chondrocytes of OA articular cartilage, which might suggest fundamental differences in the cellular physiology of these two types of hypertrophic chondrocytes. The clinical relevance of chondrocyte hypertrophy in OA has been corroborated by a study showing that cartilage calcification correlated with clinical OA scores and type X collagen expression [20].

The presence of hypertrophic chondrocytes is further supported by the presence of apoptosis markers in OA chondrocytes. It has been suggested that chondrocyte death correlates with age and severity of OA. Supporting the role of increased cell death as an integral feature of OA pathology, chondrocytes in human OA cartilage have demonstrated elevated TUNEL staining and express the apoptosis markers annexin II and V, caspase-3 and -9, BCL-2 and FAS [21]. Another theory suggests that because apoptotic cells are not removed effectively from cartilage, the products of cell death such as pyrophosphate and precipitated calcium may contribute to pathologic cartilage degradation [17]. However, it still remains controversial weather conventional apoptosis pathways decisively contribute to the pathology of OA.

Matrix Degradation

Loss of cartilage matrix is the main feature of OA. Although the exact mechanisms that underlie the onset of OA are still unknown, it is likely that a complex interplay of biomechanical and biochemical factors drives the dysregulation of chondrocyte function. The resulting imbalance between anabolic and catabolic processes in chondrocytes promotes the overexpression of matrix-degrading enzymes resulting in a loss of matrix, in particular, at the cartilage surface. The predominant proteinases responsible for the characteristic matrix degradation in OA are MMPs and aggrecanases. Among the MMPs, collagenases (MMP-1, MMP-8 and MMP-13), stromelysins (MMP-3, MMP-10, MMP-11) and gelatinases (MMP-2, MMP-9) are considered as those with the highest impact on OA cartilage breakdown [22]. In particular, MMP-13 is the enzyme responsible for most of the collagen degradation as it preferentially hydrolyses type II collagen and is about ten times more active on this molecule than MMP-1 and MMP-8 [15]. Cartilage homeostasis in normal cartilage is maintained by tissue inhibitors of matrix metalloproteinases (TIMPs) which regulate the proteolytic activity of MMPs. During OA progression, an imbalance of MMP regulation towards enhanced activity is mediated by pro-inflammatory cytokines such as interleukin-1β (IL-1β) and tumour necrosis factor-α (TNF-α) and by a shift of the MMP/TIMP balance towards MMPs.

The main activity of aggrecanases is targeted against aggrecan, which is the main proteoglycan of articular cartilage. The (a disintegrin and metalloproteinase with thrombospondin motifs) ADAMTS family is of particular interest in OA, with ADAMTS-4 and ADAMTS-5 being most active against aggrecan [23]. Similarly to MMPs, ADAMTS-4 expression is stimulated by pro-inflammatory cytokines IL-1β and TNF-α, whereas ADAMTS-5 is constitutively expressed in chondrocytes. MMPs and ADAMTSs act together to disrupt the network of collagen fibrils and proteoglycans in articular cartilage, which will eventually lead to destabilization of the joint.

Modulation of MMPs as cartilage-degrading enzymes has been a promising strategy to gain disease modification. As a consequence, a number of small molecules has been developed that can act as selective inhibitors of proteolytic MMP activity. Unfortunately, most MMP inhibitor trials have yielded disappointing results, both in orthopaedics and other disciplines, indicating that biologic systems are sometimes much more complex than predicted.

Cytokines

Many of the OA-related tissue alterations can be at least partly explained by the action of pro-inflammatory cytokines, which promote changes of chondrocyte metabolism and phenotype and lead to alterations of cartilage matrix structure. Among the cytokines believed to play a role in the progression of OA, IL-1β and TNF-α are the predominant mediators of inflammation. IL-1β and TNF-α induce MMPs and aggrecanases, reduce the synthesis of TIMPs and maintain/promote the inflammatory process via stimulation of IL-8, IL-6, nitric oxide and prostaglandin E2 production.

The impact of inflammation on the catabolic imbalance in OA cartilage suggests that it might be of therapeutic value to interfere with its major mediators, IL-1 and TNF-α. Evidence from studies on rheumatoid arthritis raised hopes that the use of IL-1 and TNF-α antagonists might prevent degradation and alterations of articular cartilage matrix in OA as well. However, despite some favourable evidence from in vitro and animal studies, inhibition of IL-1 via IL-1 receptor antagonists turned out to be of limited effectiveness in clinical OA without improving the disease progress in the long run. Similarly, clinical trials on the use of TNF-α antibodies (which are effective drugs in RA) showed that these drugs caused only a modest, not significant symptom improvement in OA.

12.2 Triggers of OA Biology

The following section aims to provide an overview of current «hot topics» in OA biology research. In particular, the selection of discussed issues covers contributing factors to OA onset and progression that might offer targets for therapeutic application in the future.

12.2.1 Ageing

Age is one of the most prominent risk factors for the development of OA. As such, the majority of people over the age of 65–70 has some radiographic signs of OA, although these signs do not necessarily correlate with symptomatic disorders. A cohort study of 90 year olds from the Netherlands found that only 16% of this aged population were free of radiographic OA [24].

The mechanisms underlying the link between age and OA are not well understood. At the cellular level, senescence can be divided into two main categories: replicative and secretory. Human articular chondrocytes have a limited proliferative capacity but a high capacity to synthesize soluble mediators. Therefore, secretory senescence should be predominant in ageing chondrocytes. This condition has been defined as the senescence-associated secretory phenotype (SASP). The SASP may negatively affect the cells' local environment by the production of cytokines and MMPs including IL-1β, IL-6, IL-8, MMP-3 and MMP-13. Mitochondrial dysfunction with increased production of ROS constitutes another hallmark of cell senescence in the articular cartilage, which also promotes chondrocyte death in OA. Thus, the SASP is a pro-inflammatory phenotype and includes many of the essential characteristics of the OA chondrocyte phenotype.

Autophagy is a cellular process used to degrade and recycle dysfunctional proteins and other macromolecules as well as to provide cells with alternative energy sources when nutrients are scarce [25]. Autophagy tends to decline with age in many cells and tissues including articular cartilage where autophagy markers were found to be decreased in OA chondrocytes [26]. A loss in autophagy was also associated with increased cell death in articular cartilage. In addition, the response of cells to growth factor stimulation is altered in senescent cells, potentially resulting in reduced repair response. Prominent examples for growth factors to which OA chondrocytes are less responsive are transforming growth factor-β (TGF-β) and insulin-like growth factor-1 (IGF-1). The loss in responsiveness to growth factor stimulation can result from oxidative stress and the resulting increased levels of ROS.

Besides the age-related changes in cartilage and chondrocytes, ageing also affects the other joint tissues such as the muscles (sarcopenia), bone (increased remodelling and bone loss), fat (increased depots) and the nervous system (altered proprioception) [25]. Hopefully, further insights into the molecular interactions between ageing and OA will deepen our understanding of why a subgroup of ageing individuals develops OA whereas others do not.

12.2.2 Obesity

Concomitant loss of muscle mass and gain of fat mass contribute to the progression of OA, primarily due to changes in joint loading.

Supported by meta-analysis, subjects that are overweight or obese have an approximately threefold higher risk of developing knee OA than those of normal weight [27]. Risk of incident knee OA increases with body mass index (BMI), and increased BMI is positively correlated with OA development in older adults and positively correlated with OA risk in younger populations. In agreement, reducing the BMI lowers the risk of developing OA, demonstrating the importance of weight control as a primary means of OA prevention. Moreover, long-term exercise and weight loss intervention was shown to significantly reduce weight joint loads in older adults. However, it has to be taken into account that the BMI is a measure of overall body weight and does not discriminate between adipose tissue and lean mass. In fact, fat and lean mass are fundamentally different regarding their metabolic activity. The observation that the risk of radiographic and symptomatic hand OA is increased twofold in obese patients supports the role of metabolic factors associated with obesity [28]. This increased risk cannot be explained with biomechanical factors but rather with systemic factors released by adipose tissue. Adipose tissue is known as a source of adipokines such as adiponectin, leptin and resistin, which contribute to the metabolic syndrome and have pro- and/or anti-inflammatory properties in OA. In agreement, adipokine blood concentrations are associated with OA severity and synovitis. Very recently, it was shown that almost half of the association between elevated BMI and knee OA could be explained by the inflammatory adipokine leptin [29]. Specifically, leptin may play an important role in the pathogenesis of OA by modulating chondrocyte functions and by contributing to osteophyte formation [30].

12.2.3 Genetics

OA has a strong genetic component. The genetic influence of this disease is estimated between 35% and 65%. The first studies performed in this context focused on the risk of twin sisters to develop OA-associated symptoms and suggested a relationship between the disease and family ties. Epidemiological studies estimate that there is a 40% probability of inheritability in an OA knee and a 65% probability of inheritability in OA hands and hips [31]. The genetics of OA are complex, as they do not follow the typical pattern of Mendelian inheritance. More likely, onset and progression of OA are associated with multiple gene interactions, supporting the concept of polygenic inheritance rather than a defect in a single gene.

Recent advances in the field of molecular genetics have made a contribution towards the identification of risk alleles for OA. Genome-wide association studies (GWAS) – that are comprehensive genomic scans searching for genes associated in some way with the pathology – have identified several loci harbouring genetic variation influencing the susceptibility to OA in humans. So far, three GWAS have been performed that combine the key requirements of extensive coverage, large cohorts and replication in additional cohorts (arcOGEN, Rotterdam study, Tokyo study). In more than 17,000 individuals (Europeans) of the arcOGEN study, five genome-wide significant loci were identified for association with OA [32]. The strongest association was with rs6976, a single nucleotide polymorphism (SNP) encoding a missense polymorphism within the nucleostemin-encoding gene GNL3.

Other candidate genes that have been reported as harbouring such risk alleles include GDF5, with the rs143383 SNP showing association in both Europeans and Asians. GDF5 encodes for a growth factor that is part of the TGF-β superfamily and is required for joint formation and maintenance. Besides its role in OA, the associated polymorphism, SNP rs143383, appears to be involved in skeletal health in general as it has been reported to be associated with a variety of other musculoskeletal phenotypes.

Other candidates include genes with a well-known functional role in the osteoarthritic process such as genes that code for structural proteins of the ECM of the cartilage, in particular those that code for type II collagen. Alterations in the COL2A1 gene might impair proper function of this most abundant protein of cartilage matrix, although a final conclusion on their clinical relevance has not yet been drawn. In addition, it is reasonable to assume that various genes, which code for different interleukins, may influence genetic susceptibility to OA. Various alterations in genes coding for IL-1, IL-1 receptor antagonist, IL-4 receptor and IL-6 have been described and related with OA [31].

Interestingly, OA risk alleles often show ethnic stratification, with associations within ethnic groups but not typically across them. This can be attributed to differences between ethnic groups in (1) the frequencies of the risk alleles, (2) the genetic background on which the risk alleles are operating and (3) non-genetic (environmental) factors that modulate the impact of the risk alleles [33].

Finally, a study on people in Spain found that the risk of OA development and progression was linked with certain mitochondrial DNA haplotypes. Whereas the mitochondrial DNA haplotype J was related with a lower risk of OA and reduced MMP-13 serum levels, the mitochondrial DNA haplotype H had increased risk for OA and higher levels of MMP-3 [34].

12.2.4 Inflammation

Traditionally, OA has been considered a «wear and tear» disease of articular cartilage triggered by any process leading to aberrant biomechanical forces on one particular joint (e.g. overload, anatomical issues, fragility of cartilage matrix). Given the absence of immune cells in the synovial fluid and of systemic manifestations of inflammation, OA has not been regarded as a classical immunological arthropathy. However, external signs and symptoms of inflammation – pain, redness, heat, swelling and loss of function – are clearly evident during OA progression. Joint swelling, which is attributed to the presence of synovitis due to thickening of the synovium or effusion, is one clinical feature of OA. Infiltration of activated B cells and T lymphocytes along with overexpression of pro-inflammatory mediators has been reported during early and late OA [35]. Furthermore, systemic high-sensitivity C-reactive protein levels reflect synovial inflammation in OA patients and are associated with the level of pain [36]. In fact, the presence of OA synovitis can be visualized using gadolinium-enhanced MRI, ultrasonography or arthroscopy and therefore may even be a surrogate marker of OA severity.

The question why the synovium becomes inflamed in OA remains controversial. The most accepted hypothesis is that, once degraded, cartilage fragments are washed out into the joint and contact the synovium. Considered foreign bodies, synovial cells react by producing inflammatory mediators, which in turn fuel catabolic activities of chondrocytes present in the superficial layer of cartilage. This leads to MMP synthesis and, eventually, degradation of the cartilage ECM. The mediators can also enhance synovial angiogenesis and increase the synthesis of inflammatory cytokines and MMPs by synovial cells themselves (vicious circle) [28]

Discussion might continue whether synovitis is an actor or a bystander in OA. However, although the essential role of synovial inflammation in OA remains controversial, it is now accepted that the inflamed synovium at least promotes cartilage degeneration and perpetuates disease progression. In chondrocytes, inflammatory processes – which were recently shown to be triggered by glycan-mediated cell surface binding of galectin-1 [37] and galectin-3 [38] – directly affect cartilage integrity via the overexpression of MMPs.

The innate immune system comprises the cells and mechanisms that defend the host from infection by other organisms in a non-specific manner. Toll-like receptors (TLRs) are the basic signalling receptors of the innate immune system. Interestingly, the expression of TLR-2 and TLR-4 was shown to be up-regulated in lesional areas of OA cartilage. Concomitantly, ligands of TLR-2 and TLR-4 such as low-molecular-weight hyaluronic acid, fibronectin, tenascin-C and alarmins (S100 proteins, high-mobility group protein B1 (HMGB1)) have been found in OA synovial fluid [28, 39]. Of note, these factors can induce catabolic responses in chondrocytes and/or inflammatory responses in synoviocytes.

Inflammatory events occurring within joint tissues could be reflected outside the joint in plasma and peripheral blood leukocytes (PBLs) of patients with OA. Levels of several inflammatory mediators are higher in OA than healthy sera. In the Leiden 85-plus study, 16% of 90-year-old people were «free from OA». Blood samples taken from these persons were treated with LPS to measure the cytokine production capacity of the innate immune system. Interestingly, subjects in the lowest tertile of Il-1β production had an 11-fold increased chance to be free of OA [40]. In agreement, another study assessed gene expression profiles in PBLs from patients with OA and revealed two distinct subgroups: one with increased level of IL1B and one with normal expression. Interestingly, patients with the inflammatory «IL1B signature» had higher pain scores and decreased function and were at higher risk of radiographic progression of OA [41].

Recently, the concept of inflammaging has been introduced to the biology of OA [42]. The term inflammaging describes the interplay between chronic inflammation and age, which nowadays is accepted to play an important role in the initiation and/or progression of age-related diseases including OA. It has been recognized that senescent cells represent a source of chronic inflammation, characterized by the continued presence of pro-inflammatory factors at levels higher than baseline but manyfold lower than those found in acute inflammation. Among the dozens of factors, IL-6, IL-8, IL-1, monocyte chemotactic protein (MCP)-2, MCP-3, MMP-1, MMP-3 and many of the IGF-binding proteins contribute to the SASP. Most SASP proteins are up-regulated primarily at the transcriptional level by the transcription factors NF-κB and C/EBPβ, which have increased activity in senescent cells. The activities of C/EBPβ and NF-κB are regulated by multiple pathways depending on the cellular context, and the signalling pathways that drive the SASP are just beginning to emerge. However, there is evidence that genotoxic stress induced, e.g. by ionizing radiation, hyperproliferation due to oncogene activity or dysfunctional telomeres, promotes the development of SASP.

Moreover, the positive feedback loop maintained by IL-1α sustains the SASP, reinforcing its expression and the senescence growth arrest, also via NF-κB and C/EBPβ activity. MicroRNAs also contribute to the regulation of the SASP. So far, two microRNAs, miR-146a and miR-146b, have been shown to negatively regulate the senescence-associated secretion of IL-6 and IL-8 by targeting IRAK1, which is a positive regulator of NF-κB. In addition, senescent cells undergo global changes in chromatin organization, and it is probable that these changes have an effect on SASP gene regulation.

Although it remains to be proven that senescent cells and the SASP are drivers of normal ageing and age-related diseases, further advances in this field might hopefully facilitate the development of therapies aimed at mitigating the deleterious effects of the SASP, however, without compromising its beneficial effects (e.g. tissue renewal upon tissue damage).

12.2.5 Subchondral Bone

The subchondral bone is the zone lying immediately underneath the calcified zone of the articular cartilage. The subchondral bone is separated from the calcified zone by the «cement line» and consists of the subchondral bone plate and the subarticular spongiosa. The subchondral plate is an integral and dynamic component of the joint as it provides support for the overlying articular cartilage and absorbs most of the mechanical force transmitted by joints. However, the subchondral bone cannot be reduced to its shock absorbing qualities; it also plays an important role in articular cartilage metabolism [43].

Under OA conditions, the subchondral bone undergoes a series of pathological alterations including sclerosis with thickening of cortical plate, extensive remodelling of the trabeculae, formation of osteophytes and the development of subchondral bone cysts [44, 45]. Osteoarthritic subchondral bone osteoblasts are characterized by enhanced anabolic activity under OA conditions, which was estimated to be increased 20-fold compared to normal bone turnover [46]. It has been reported that subchondral bone explants from OA patients aberrantly secrete a range of biochemical factors – including alkaline phosphatase, osteocalcin, osteopontin, IL-6, IL-8, prostaglandin, IGF-1, IGF-2 and TGF-β – that might contribute to abnormal bone remodelling [47].

The functional interplay between articular cartilage and subchondral bone led to the concept of a biochemical and molecular crosstalk across these joint-forming tissues. As such, a study on bovine explants demonstrated that the subchondral bone significantly influenced chondrocyte survival in vitro [48].

This concept would also suggest that specific cellular and molecular interactions might lead to the progression of OA. The presence of microcracks in the OA bone matrix might allow the direct interaction between chondrocytes and subchondral bone cells via secreted mediators, also mediated by the synovial fluid. Similarly, increased vascularity within the deep layers of articular cartilage facilitates molecular transport through the calcified tissues [45]. Via these routes, chemokines, cytokines and proteases secreted from chondrocytes could reach subchondral bone osteoblasts to modify their biochemical and functional characteristics. For example, IL-6 in combination with IL-1β can switch osteoblasts from a normal to a sclerotic phenotype [49]. Moreover, subchondral bone loss in OA can be promoted by chondrocyte-secreted regulatory factors that induce osteoclastogenesis [50].

The clinical relevance of the interaction between cartilage and subchondral bone has been addressed in animal models and clinical studies suggesting that pathological remodelling of the subchondral bone in OA may actually precede and mediate cartilage degeneration [51, 52]. Thus, targeting the crosstalk between OA cartilage and bone tissues with drugs, e.g. bisphosphonates, strontium ranelate or cathepsin K inhibitors, may represent an attractive therapeutic approach [52].

Taken together, the subchondral bone may have a substantial role in the OA process, as a mechanical damper, as well as a source of inflammatory mediators implicated in the OA pain process and in the degradation of cartilage [28]. Even though the crosstalk between articular cartilage and subchondral bone appears reasonable, the primacy of cartilage versus subchondral bone alteration for the onset of OA cartilage degeneration is still under debate. It is anticipated that a better understanding of the basic science and pathophysiology of the subchondral region will translate into improved strategies for the repair of cartilage defects that arise from or extend into the subchondral bone [43].

12.2.6 Signalling

In healthy joints, balance between various anabolic and catabolic signalling pathways is required to maintain functionality of articular cartilage and other joint tissues. An imbalance of this delicate equilibrium, due to a number of reasons as discussed in above sections, leads to gradual deterioration of cartilage quality contributing to the progression of OA.

In the following, a few of the signalling pathways that have been implicated in OA pathogenesis will be presented.

NF-κB

The NF-κB pathway is a central regulator of catabolic actions in OA, mediating inflammatory responses of chondrocytes and leading to ECM damage and cartilage erosion. The NF-κB pathway, which is induced by galectin-1 and galectin-3 in OA chondrocytes [37, 38], is essential for chondrocytes to express inflammation-related genes coding for IL-1, IL-6, TNF-α, cyclooxygenase-2, inducible nitric oxide synthase and MMP-1, MMP-3 and MMP-13. Besides activation of inflammation, NF-κB transcription factors (1) trigger the differentiation programme of chondrocytes towards a more differentiated, hypertrophic-like phenotype, (2) participate in the chondrocyte catabolic responses to ECM degradation products and mechanical stress and (3) play important roles in both positive and negative regulation of the SASP.

NF-κB-mediated transcriptional control arises from the assembly of homodimers and heterodimers of five different NF-κB proteins (RelA/p65, RelB, c-Rel, NFκB1/p105 and NFκB2/p100). NF-κB dimers are sequestered in the cytoplasm, and their transcriptional activities are blocked by one of three small inhibitory NF-κB proteins (IκBα, IκBβ, IκBε) or the larger precursor proteins p105 (NFκB1) and p100 (NFκB2). Upon stimulation, the canonical NF-κB heterodimers (predominantly p65-p50) are activated by phosphorylation of IκBα by the IKK (inhibitor of NF-κB kinase) complex that targets it for ubiquitination and subsequent proteasomal destruction. The activated NF-κB heterodimers are then translocated into the nucleus and can activate their specific target genes [53].

The NF-κB signalling pathway provides multiple avenues for targeting OA. Although still in their infancy, strategies to therapeutically target components of the NF-κB pathway have been introduced in recent years. This includes inhibition of IKK-mediated phosphorylation, IκBα ubiquitination, nuclear translocation or transcriptional induction of NF-κB target genes using anti-inflammatory compounds (synthetic or natural products), siRNAs or decoy oligodeoxynucleotides [53].

WNT

The wingless-type (WNT) signalling pathways consists of at least 19 structurally related glycoproteins which can transduce their signal through different intracellular cascades [45]. The canonical WNTs (e.g. WNT-1, WNT-3a, WNT-8) are characterized by their ability to promote translocation of β-catenin to the nucleus, whereas the non-canonical WNTs (e.g. WNT-4, WNT-5a, WNT-11) do not affect β-catenin levels. β-Catenin is not itself a transcription factor but binds to a number of transcription factors, co-activators or co-repressors. The WNT signalling cascades have essential roles in development, growth and homeostasis of joints and the skeleton [54]. In cartilage, canonical WNT signalling is necessary for the maintenance of the functional articular cartilage phenotype, which is characterized by extended cell survival and absence of differentiation towards hypertrophy. Of note, overexpression of WNT signalling can be deleterious to chondrocytes, leading to OA-like alterations. During the progression of OA, differential expression of various WNT antagonists (e.g. gremlin-1, Frizzled-related protein (FRZB)-3) and agonists (e.g. WNT-16) was demonstrated in the cartilage, bone and synovium. The aberrant secretion of agonists in joints may directly stimulate chondrocytes to secrete increased MMPs and aggrecanases to enhance deterioration of cartilage, whereas the subchondral bone may respond with osteophyte formation [55].

Given their roles in joint biology, members of the WNT signalling cascades have been considered as potential biomarkers to assess skeletal tissue turnover and disease progression. Recently, the WNT signalling antagonists FRZB and Dickkopf-related protein 1 were shown to provide prognostic information about OA development and progression in patients with radiographic hip OA [56]. The complexity of WNT signalling represents a tremendous challenge for the therapeutic modulation of this pathway. Despite recent advances in our understanding of the role of WNT pathway in joint pathophysiology, open questions regarding the source, timely involvement and precise mechanism of action of the diverse WNT factors will have to be answered before therapeutics that target WNT signalling can be developed [45].

TGF-β/BMP

The TGF-β family of cytokines, including TGF-β and bone morphogenetic proteins (BMPs), plays critical roles in embryonic development, adult tissue homeostasis and the pathogenesis of a variety of diseases. BMP-2, BMP-4 and BMP-5 are required for chondrocyte proliferation and matrix synthesis via canonical Smad molecules [15]. The involvement of locally secreted BMPs in cartilage biology is not yet clear, but a few studies have shown that BMPs play significant roles in protection and repair of cartilage by regulating the synthesis of aggrecan and proteoglycans. However, BMPs are also required for the chondrocyte hypertrophy in OA, indicating that BMP signalling not only plays important roles in the early stages of chondrogenesis by stimulating the synthesis of matrix molecules like type II collagen but also participates during terminal differentiation by elevating MMP-13 expression, as observed in OA cartilage [45].

TGF-β signalling might also be a mediator of the crosstalk between articular cartilage and subchondral bone, as inhibition of TGF-β1 activity in subchondral bone led to less degeneration of articular cartilage in different rodent models. Similarly, increased TGF-β levels in subchondral bone induced OA-like symptoms in mice, including lower expression of proteoglycan, increased thickness of calcified cartilage and increased angiogenesis in subchondral plate [45].

MAPK

The family of mitogen-activated protein kinases (MAPKs) comprises three broad categories of kinases: extracellular signal-regulated kinases (ERKs), stress-activated protein kinases/c-Jun N-terminal kinases (JNKs) and p38 kinases [45]. MAPKs have a critical role in both cartilage and bone biology, as they appear to be involved in the transduction of mechanical signals in cartilage development. In the bone, ERK, p38 and JNK promote osteoclast differentiation by regulating activator protein-1 (AP-1) as a critical mediator of osteoclastogenesis. MAPKs family has been suggested to be involved in the pathophysiology of OA. ERK and p38 activation are key upstream signalling events in processes leading to degeneration of articular cartilage. Activation of both ERK and p38 signalling is essential for MMP expression and activity, while only ERK activation is essential for aggrecanase-mediated cartilage degradation [57]. The possibility of MAPK-mediated release of degradative enzymes from the subchondral bone affecting chondrocytes suggests the existence of intercellular communication between the cartilage and bone affecting each other. In fact, co-culture of normal or OA subchondral bone with articular cartilage elucidated an intercellular crosstalk between both tissues mediated by MAPKs [58]. Another study demonstrated that OA subchondral osteoblasts stimulate hypertrophic gene expression and matrix calcification in articular chondrocytes by inducing ERK signalling activity and downregulating p38 signalling activity [59]. However, further studies are needed to translate our understanding of the intercellular communication between articular cartilage and subchondral bone into therapeutic interventions.

12.3 Concluding Remarks

Given the socioeconomic impact and the multimodality of OA as well as the unique properties of joint tissues, OA biology represents a challenging field of medical research. Understanding the basic science of joint tissues and the changes that occur in OA is imperative to develop novel strategies to diagnose and treat this disorder.

One central question of upcoming research efforts in the field of OA biology might be: What keeps an articular chondrocyte functioning as an articular chondrocyte, or vice versa, what prompts articular chondrocytes to fail in maintaining the functionality of cartilage? [60].

Even more fundamental is the question as to the upstream mechanisms of inflammation and matrix degradation, which might offer novel therapeutic targets. In this context, scientific progress might also shed light on the uncertain role of cartilage in the disease process: Are chondrocytes main actors in cartilage pathology or just passive responders to disease processes initiated elsewhere? And what would be the consequences for therapeutic strategies?

In summary, a deeper understanding of the molecular mechanisms maintaining chondrocyte homeostasis, also with respect to the crosstalk with other joint tissues, will be mandatory to achieve scientific and biomedical progress in orthopaedic research. Although articular cartilage is the tissue considered to be mainly affected by OA and chondrocytes are the major players in cartilage degradation, other joint tissues as well as systemic metabolic processes determine the development and progression of OA. This view of OA, that has changed from a single tissue disease to a «complex, multisystemic disease», will hopefully provide several new levels that can be targeted for the successful treatment of OA patients.

> **Take-Home Message**
> - OA is a degenerative disease of the whole joint, with enormous socioeconomic impact and of unclear aetiology.
> - Pathophysiologic hallmarks of OA include the degradation of cartilage extracellular matrix, pro-inflammatory events in synovium and cartilage and the phenotypic modulation of chondrocytes.
> - Ageing, obesity and genetic factors can drive OA onset and progression.
> - The imbalance between anabolic and catabolic processes in joint tissues is promoted by the malfunction of signalling pathways such as the NF-κB, Wnt, TGF-β/BMP or MAPK cascades.
> - Concerted research efforts appear necessary to elucidate the precise molecular mechanisms and upstream effectors, with the major goal of developing novel and effective OA therapies.

References

1. Glyn-Jones S, et al. Osteoarthritis. Lancet. 2015;386:376–87.
2. Mehrnaz Maleki-Fischbach JMJ. New developments in osteoarthritis. Sex differences in magnetic resonance imaging-based biomarkers and in those of joint metabolism. Arthritis Res Ther. 2010;12:212.
3. Cattano NM, et al. Joint trauma initiates knee osteoarthritis through biochemical and biomechanical processes and interactions. OA Musculoskeletal Medicine. 2013;1:3.
4. Neogi T, Zhang Y. Epidemiology of osteoarthritis. Rheum Dis Clin N Am. 2013;39:1–19.
5. Hoeven TA, et al. Association of atherosclerosis with presence and progression of osteoarthritis: the Rotterdam study. Ann Rheum Dis. 2013;72:646–51.
6. Woolf AD, Pfleger B. Burden of major musculoskeletal conditions. Bull World Health Organ. 2003;81:646–56.
7. Rabenberg M. Arthrose. Gesundheitsberichterstattung des Bundes. 2013;54:1–36.
8. Wieland HA, Michaelis M, Kirschbaum BJ, Rudolphi KA. Osteoarthritis – an untreatable disease? Nat Rev Drug Discov. 2005;4:331–44.
9. Palazzo C, Ravaud J-F, Papelard A, Ravaud P, Poiraudeau S. The burden of musculoskeletal conditions. PLoS One. 2014;9:e90633.
10. Bachinger E. Vienna health report 2010. City of Vienna, Vienna, 2010.

11. United Nations, D. O. E., Social Affairs. P. D. World population prospects: the 2012 revision. United Nations New York, 2013.
12. Turkiewicz A, Petersson IF, Björk J, Dahlberg LE, Englund M. The consultation prevalence of osteoarthritis 2030 may increase by 50%: prognosis for Sweden. Osteoarthr. Cartilage. 2013;21:S160–1.
13. Pearle AD, Warren RF, Rodeo SA. Basic science of articular cartilage and osteoarthritis. Clin Sports Med. 2005;24:1–12.
14. Goldring MB, Marcu KB. Cartilage homeostasis in health and rheumatic diseases. Arthritis Res Ther. 2009;11(2):224.
15. Umlauf D, Frank S, Pap T, Bertrand J. Cartilage biology, pathology, and repair. Cell Mol Life Sci. 2010;67:4197–211.
16. Sellam J, Berenbaum F. The role of synovitis in pathophysiology and clinical symptoms of osteoarthritis. Nat Rev Rheumatol. 2010;6:625–35.
17. Sandell LJ, Aigner T. Articular cartilage and changes in arthritis. An introduction: cell biology of osteoarthritis. Arthritis Res. 2001;3:107–13.
18. Drissi H, Zuscik M, Rosier R, O'Keefe R. Transcriptional regulation of chondrocyte maturation: potential involvement of transcription factors in OA pathogenesis. Mol Asp Med. 2005;26:169–79.
19. Pitsillides AA, Beier F. Cartilage biology in osteoarthritis —lessons from developmental biology. Nat Rev Rheumatol. 2011;7:654–63.
20. Fuerst M, et al. Calcification of articular cartilage in human osteoarthritis. Arthritis Rheum. 2009;60:2694–703.
21. van der Kraan PM, van den Berg WB. Chondrocyte hypertrophy and osteoarthritis: role in initiation and progression of cartilage degeneration? Osteoarthr Cartil. 2012;20:223–32.
22. Burrage PS, Mix KS, Brinckerhoff CE. Matrix metalloproteinases: role in arthritis. Front Biosci. 2006;11:529–43.
23. Song R-H, et al. Aggrecan degradation in human articular cartilage explants is mediated by both ADAMTS-4 and ADAMTS-5. Arthritis Rheum. 2007;56:575–85.
24. Goekoop RJ, et al. Determinants of absence of osteoarthritis in old age. Scand J Rheumatol. 2011;40:68–73.
25. Loeser RF. Aging and osteoarthritis. Curr Opin Rheumatol. 2011;23:492–6.
26. Caramés B, Taniguchi N, Otsuki S, Blanco FJ, Lotz M. Autophagy is a protective mechanism in normal cartilage, and its aging-related loss is linked with cell death and osteoarthritis. Arthritis Rheum. 2010;62:791–801.
27. Blagojevic M, Jinks C, Jeffery A, Jordan KP. Risk factors for onset of osteoarthritis of the knee in older adults: a systematic review and meta-analysis. Osteoarthr Cartil. 2010;18:24–33.
28. Berenbaum F. Osteoarthritis as an inflammatory disease (osteoarthritis is not osteoarthrosis!). Osteoarthr. Cartilage. 2013;21:16–21.
29. Fowler-Brown A, Kim DH, Shi L, Marcantonio E, Wee CC, Shmerling RH, Leveille S. The mediating effect of leptin on the relationship between body weight and knee osteoarthritis in older adults. Arthritis Res. 2015;67(1):169–75.
30. Dumond H, et al. Evidence for a key role of leptin in osteoarthritis. Arthritis Rheum. 2003;48:3118–29.
31. Fernández-Moreno M, Rego I, Carreira-Garcia V, Blanco FJ. Genetics in osteoarthritis. Curr Genomics. 2008;9:542–7.
32. Consortium A, Collaborators A. Identification of new susceptibility loci for osteoarthritis (arcOGEN): a genome-wide association study. Lancet. 2012;380:815–23.
33. Reynard LN, Loughlin J. The genetics and functional analysis of primary osteoarthritis susceptibility. Expert Rev Mol Med. 2013;15:e2.
34. Rego-Pérez I, et al. Mitochondrial DNA haplogroups and serum levels of proteolytic enzymes in patients with osteoarthritis. Ann Rheum Dis. 2011;70:646–52.
35. Benito MJ, Veale DJ, FitzGerald O, van den Berg WB, Bresnihan B. Synovial tissue inflammation in early and late osteoarthritis. Ann Rheum Dis. 2005;64:1263–7.
36. Pearle AD, et al. Elevated high-sensitivity C-reactive protein levels are associated with local inflammatory findings in patients with osteoarthritis. Osteoarthr Cartilage. 2007;15:516–23.
37. Toegel S, et al. Galectin-1 couples glycobiology to inflammation in osteoarthritis through the activation of an NF-κB-regulated gene network. J Immunol. 2016;196:1910–21.
38. Weinmann D, et al. Galectin-3 induces a pro-degradative/inflammatory Gene signature in human chondrocytes, teaming up with Galectin-1 in osteoarthritis pathogenesis. Sci Rep. 2016;6:39112.
39. Kim HA, et al. The catabolic pathway mediated by toll-like receptors in human osteoarthritic chondrocytes. Arthritis Rheum. 2006;54:2152–63.

40. Goekoop RJ, et al. Low innate production of interleukin-1beta and interleukin-6 is associated with the absence of osteoarthritis in old age. Osteoarthr Cartil. 2010;18:942–7.
41. Attur M, et al. Increased interleukin-1β gene expression in peripheral blood leukocytes is associated with increased pain and predicts risk for progression of symptomatic knee osteoarthritis. Arthritis Rheum. 2011;63:1908–17.
42. Freund A, Orjalo AV, Desprez P-Y, Campisi J. Inflammatory networks during cellular senescence: causes and consequences. Trends Mol Med. 2010;16:238–46.
43. Madry H, van Dijk CN, Mueller-Gerbl M. The basic science of the subchondral bone. Knee Surg Sports Traumatol Arthrosc. 2010;18:419–33.
44. Thambyah A, Broom N. On new bone formation in the pre-osteoarthritic joint. Osteoarthr Cartil. 2009;17:456–63.
45. Sharma AR, Jagga S, Lee S-S, Nam J-S. Interplay between cartilage and subchondral bone contributing to pathogenesis of osteoarthritis. Int J Mol Sci. 2013;14:19805–30.
46. Bailey AJ, Mansell JP, Sims TJ, Banse X. Biochemical and mechanical properties of subchondral bone in osteoarthritis. Biorheology. 2004;41:349–58.
47. Sanchez C, et al. Phenotypic characterization of osteoblasts from the sclerotic zones of osteoarthritic subchondral bone. Arthritis Rheum. 2008;58:442–55.
48. Amin AK, Huntley JS, Simpson AHRW, Hall AC. Chondrocyte survival in articular cartilage: the influence of subchondral bone in a bovine model. J Bone Joint Surg Br. 2009;91:691–9.
49. Sanchez C, et al. Osteoblasts from the sclerotic subchondral bone downregulate aggrecan but upregulate metalloproteinases expression by chondrocytes. This effect is mimicked by interleukin-6, −1beta and oncostatin M pre-treated non-sclerotic osteoblasts. Osteoarthr Cartil. 2005;13:979–87.
50. Jiao K, et al. Subchondral bone loss following orthodontically induced cartilage degradation in the mandibular condyles of rats. Bone. 2011;48:362–71.
51. Neogi T. Clinical significance of bone changes in osteoarthritis. Ther Adv Musculoskelet Dis. 2012;4:259–67.
52. Tonge DP, Pearson MJ, Jones SW. The hallmarks of osteoarthritis and the potential to develop personalised disease-modifying pharmacological therapeutics. Osteoarthr Cartil. 2014;22(5):609–21.
53. Marcu KB, Otero M, Olivotto E, Borzi RM, Goldring MB. NF-kappaB signaling: multiple angles to target OA. Curr Drug Targets. 2010;11:599–613.
54. Lories RJ, Corr M, Lane NE. To Wnt or not to Wnt: the bone and joint health dilemma. Nat Rev Rheumatol. 2013;9:328–39.
55. Blom AB, et al. Involvement of the Wnt signaling pathway in experimental and human osteoarthritis: prominent role of Wnt-induced signaling protein 1. Arthritis Rheum. 2009;60:501–12.
56. Lane NE, et al. Wnt signaling antagonists are potential prognostic biomarkers for the progression of radiographic hip osteoarthritis in elderly Caucasian women. Arthritis Rheum. 2007;56:3319–25.
57. Sondergaard B-C, et al. MAPKs are essential upstream signaling pathways in proteolytic cartilage degradation--divergence in pathways leading to aggrecanase and MMP-mediated articular cartilage degradation. Osteoarthr Cartil. 2010;18:279–88.
58. Prasadam I, et al. Osteoarthritic cartilage chondrocytes alter subchondral bone osteoblast differentiation via MAPK signalling pathway involving ERK1/2. Bone. 2010;46:226–35.
59. Prasadam I, et al. ERK-1/2 and p38 in the regulation of hypertrophic changes of normal articular cartilage chondrocytes induced by osteoarthritic subchondral osteoblasts. Arthritis Rheum. 2010;62: 1349–60.
60. van der Kraan PM. Osteoarthritis year 2012 in review: biology. Osteoarthr Cartil. 2012;20:1447–50.

Supplementary Information

Index – 207

Index

A

Acute joint inflammation 79
Adventitial reticular cells (ARCs) 18
Algodystrophy 80, 81
Alkaline phosphatase (AP) 20, 59
Amino-terminal (NTX-1) 58
Anti-resorptive therapy 63
Arthritis 180, 181
Articular cartilage 140, 192
– characteristics 142
– chondrocyte and pericellular microenvironment 148
– electron microscopic image of 143
– function 141
– morphology 141–143
– structure 149–150
– subdivision of 143–146
ATDC5 166
Autophagy 194

B

B cell depletion 183
Bisphosphonate(s) 43
Bisphosphonate-related osteonecrosis of the jaw (BRONJ) 129, 130
BMP. See Bone morphogenetic proteins (BMPs)
Bone 84–86
– formation markers
  – alkaline phosphatase 59
  – osteocalcin 60
  – type 1 collagen 59
– fragility
  – bone mass 84
  – cortical microarchitecture 85
  – macroarchitecture 85
  – material property 86
  – microdamage 86
  – trabecular microarchitecture 85
  – turnover 86
Bone material 2, 4, 5, 7, 8
– characterization 8–11
– fragility 12
– mechanical properties 6
– osteogenesis imperfecta 12
– postmenopausal osteoporosis 12
– quality
  – microcomposite 7
  – mineralization pattern 7
  – toughening mechanisms 8

– structure and composition
  – BSUs 2
  – collagen and mineral 5
  – collagen fibrils arrangement 4
  – hierarchical level 2
  – mineralized collagen fibril 5
Bone metastases 40
Bone mineral density (BMD) 84
Bone morphogenetic proteins (BMPs) 131–133
Bone resorption 36, 58, 59
Bone scintigraphy 74, 75, 77
– definition 68
– inflammation 78
– metabolic bone disorders
  – fibrous dysplasia 77
  – hyperthyroidism 77
  – osteomalacia 75, 76
  – osteoporosis 74
  – Paget's disease 77, 78
  – primary hyperparathyroidism 75, 76
  – renal osteodystrophy 77
– metastases 72
– phases of investigation 69
– primary tumors 70
Bone structural units (BSUs) 2–3
Bone turnover markers (BTM)
– animal study 60
– cathepsin K (Cat K) inhibitors 61
– clinical relevance 63
– Dkk1 61
– FGF-23 62
– miRNAs 62
– non-pathological and pathological conditions 62
– osteoporosis treatment 63
– PSTN 61
– RANK/RANKL/OPG system 61
– sclerostin 61
BSUs. See Bone Structural Units (BSUs)
BTM. See Bone turnover markers (BTM)

C

Carboxy-terminal (CTX-1) 58
Cartilage. See also Articular cartilage
– research morphological methods for 150
– types 140
Cathepsin K (Cat K) inhibitors 44, 61
Chondrocytes 147, 148

– media conditions 167
– morphology and arrangement of 144
– phenotypic modulation of 192
Chronic kidney disease (CKD) 62
Chronic kidney disease-mineral bone disorder (CKD-MBD) 94
Chronic osteomyelitis 78, 79
CKD. See Chronic kidney disease (CKD)
CKD-MBD. See Chronic kidney disease-mineral bone disorder (CKD-MBD)
Cleft lip 127
Collagen type 1 58
Cortical microarchitecture 85
Cytokines 194

D

Dendritic osteocytes 24
Denosumab 43
Dickkopf 1 (Dkk1) 61
Dual-energy x-ray absorptiometry (DXA) 57

E

Electron microscopy 161
Endochondral ossification 154
Extracellular matrix (ECM)
– compartmentalisation 147
– structural organisation 142

F

Facial skeleton 126
FGF-23. See Fibroblast growth factor 23 (FGF-23)
Fibroblastic growth factor(s) (FGFs) 157
Fibroblast growth factor-23 (FGF-23) 62, 94
Fibrous cartilage 140
Fibrous dysplasia 77
Fourier transform infrared (FTIR) 9

G

Gene expression analysis 163
Growth plates (GP) 154–158
– electron microscopy 161
– histology 158

Growth plates (GP) (cont.)
- immunohistochemistry 160
- in vitro model 165
- LCM 163–164
- structure and function
  - differentiation and proliferation 155
  - endochondral ossification 154
  - endocrine regulation 157
  - FGF 157
  - mechanism 158
  - PTHrPR 157
  - SOX9 156
- transgenic animal models 164–165

H

Haematopoietic stem cells (HSCs), regulation of 38, 39
Histochemical TRAcP assay 46
HOPE® fixation 159
Human peripheral blood mononuclear cells 45, 46
Hyaline cartilage 140
Hydroxyproline 58
Hypermetabolic bone disease 76
Hyper-osmolar fixative solution 161
Hyperthyroidism 77
Hypertrophic chondrocytes 156, 193
Hypertrophy 155, 156

I

IL-1 neutralization 182
IL-6 blockade 182
In vitro model
- 3D culture systems conditions 167, 168
- chondrocyte media conditions 167
- metatarsal culture 166
- primary cells 165–166
Inflammation 78
Inflammatory joint disease 174–177, 180–183
- rheumatoid arthritis
  - B cell depletion 183
  - co-stimulation modulation 183
  - IL-1 neutralization 182
  - IL-6 blockade 182
  - TNF blockade 181, 182
- synovial inflammation 175
  - (auto-)antigen presentation 176
  - B cells, rheumatoid 176, 177
  - cytokines and mesenchymal tissue response 180–181
  - innate immune system 175, 176
  - T cells, rheumatoid 177, 180
Interterritorial matrix 147

J

Joint inflammation 78

L

Laser capture microdissection (LCM) 163–164

M

Macroarchitecture 85
- cortical 85–86
- microdamage 86
- trabecular 85
Macrophage colony-stimulating factor (M-CSF) 35, 44
Malignant infantile osteopetrosis 42
MAPK. See Mitogen-activated protein kinases (MAPK)
Mesenchymal stem cells (MSCs) 18, 19, 166
Metabolic bone disorders
- fibrous dysplasia 77
- hyperthyroidism 77
- osteomalacia 75, 76
- osteoporosis 74
- Paget's disease 77, 78
- primary hyperparathyroidism 75, 76
- renal osteodystrophy 77
Microdamage 86
MicroRNAs (miRNAs) 62
Mineral particles/crystals, characteristics 9, 10
Mitogen-activated protein kinases (MAPKs) 201

N

NF-κB pathway 199, 200
Non-osteoporotic bone diseases
- CKD-MBD 94
- Paget's disease 91
- primary hyperparathyroidism 92
- rickets and osteomalacia 92
Nuclear medicine technique 68

O

OA. See Osteoarthritis (OA)
OPG. See Osteoprotegerin (OPG)
Organic matrix, arrangement and composition 8
Osteoarthritis (OA) 190, 192–201
- biology
  - aging 194–195
  - genetic 195–196
  - inflammation 196–198
  - obesity 195
  - signalling 199–201
  - subchondral bone 198
- pathophysiology 191
  - cytokines 194
  - matrix degradation 193
  - phenotypic modulation, of chondrocytes 192–193
- risk factors 190
- socioeconomic impact 191
Osteoblasts 18
- experimental systems 25–29
- osteogenic vs.adipogenic programming 19–21
- regulation of 38
- to osteocytes 22
Osteocalcin (Oc) 21, 60
Osteochondrosis 80
Osteoclastogenesis 34
Osteoclasts 36–47
- biology 34
- differentiation 35
  - evaluation 46
- functions
  - bone resorption 36–38
  - evaluation 47
  - haematopoietic stem cells, regulation of 38–39
  - regulation of osteoblasts 38
- isolation
  - from human peripheral 45
  - from mouse bone marrow cells 45
- osteoclastogenesis 34–36
- pharmacology
  - antiresorptive agents 44
  - bisphosphonates 43
  - denosumab 43
- physiology
  - bone metastases 40–41
  - osteopetrosis 41–42
  - osteoporosis 39–40
- primary culture 44–47
Osteocyte Lacunar-Canalicular network (OLCN) 10
Osteocytes
- cells 28
- definition 22
- different transitional stages and sequential expression 24
- essential facts 23
- mesenchymal stem cells 18, 19
- osteoblast to 22
Osteogenesis imperfecta (OI) 12
Osteomalacia 75, 76, 92, 93
Osteonecrosis, medication-related 129, 131

Index

Osteonectin 20
Osteopetrosis 41
Osteoporosis 39, 57, 74, 89
– definition 87
– pathophysiology 88
– postmenopausal 12
– postmenopausal bone loss 90
– risk factors 87
– senile bone loss 90
– treatment 63
– types 88
Osteoprotegerin (OPG) 61

P

Paget's disease 77, 78, 91
Palate intubate 127
PATN. See Periostin (PSTN)
Pericellular matrix 147
Periodontitis
– classification 128
– definition 127
Periostin (PSTN) 61
Primary cell
– cell lines 166
– cell subfractioning 166
– isolation 165
– mesenchymal stem cells 166
Primary hyperparathyroidism 75, 92
Primary osteoporosis 88
Proteoglycans 20
PTHrP receptor (PTHrPR) 157
Pyridinium cross-links 58

Q

Quantitative backscattered electron imaging (qBEI) 9
Quantitative microradiography (qMR) 9

R

Raman microspectroscopy (RM) 9
RANK. See Receptor activator nuclear factor κ B (RANK)

Receptor activator nuclear factor κB ligand (RANKL) 61
Receptor activator of nuclear factor κB (RANK) 35, 36, 61
Reflex sympathetic dystrophy (RSD) 81
Renal osteodystrophy 77
Reverse transcription polymerase chain reaction (RT-PCR) 163
Rheumatoid arthritis (RA)
– B cell depletion 183
– co-stimulation modulation 183
– IL-1 neutralization 182
– IL-6 blockade 182
– TNF blockade 181
Rheumatoid inflammation
– B cells 176
– innate immune system 175
– T cells 177–180
– Th1 cell 178
– Th17 cell 179
– Th22 T cell 179
– tregs in 179
Rickets 92, 93
RNA isolation 163
RT-PCR. See Reverse transcription polymerase chain reaction (RT-PCR)
Runt-related transcription factor 2 (RUNX2) 19

S

Scanning acoustic microscopy (SAM) 11
Scanning small-angle X-ray scattering (sSAXS) 9
Scanning wide-angle X-ray scattering (sWAXS) 10
Sclerostin 61
Senile bone loss 90
Sex-determining region Y-box 9 (SOX9) 156
Split lines 147
Subchondral bone 198–199
Synchrotron radiation microtomography (SRμCT) 9
Synovial inflammation 175
– (auto-)antigen presentation 176

– B cells, rheumatoid 176
– cytokines and mesenchymal tissue response 180
– innate immune system 175
– T cells, rheumatoid 177
Systemic autoimmune disorders 174

T

Tartrate-resistant acid phosphatase (TRAP) 38, 46, 59
Tc-99m-labeled diphosphonates 69
Territorial matrix 147
3D culture systems 167
TNF blockade 181
Trabecular microarchitecture 85
TRAcP. See Tartrate-resistant acid phosphatase (TRAP)
Transforming growth factor beta (TGFβ) signalling pathway 200
Transgenic animal models
– caveats 165
– conditional gene targeting 164
– global knockouts 164
Transmission electron microscopy (TEM) 10
TRAP. See Tartrate-resistant acid phosphatase type 5 (TRAP)
Tumor-induced osteomalacia (TIO) 94
Type 1 collagen 58, 59

V

Vitamin D_3 92
αVβ3 integrin inhibitors 44

W

Wingless-type (WNT) signalling pathway 200

X

X-linked hypophosphatemic rickets (XLH) 94

If you have any concerns about our products,
you can contact us on
ProductSafety@springernature.com

In case Publisher is established outside the EU,
the EU authorized representative is:
**Springer Nature Customer Service Center GmbH
Europaplatz 3, 69115 Heidelberg, Germany**

Printed by Libri Plureos GmbH
in Hamburg, Germany